Robert Mallet

Great Neapolitan Earthquake of 1857

Vol. I

Robert Mallet

Great Neapolitan Earthquake of 1857
Vol. I

ISBN/EAN: 9783337173388

Printed in Europe, USA, Canada, Australia, Japan

Cover: Foto ©ninafisch / pixelio.de

More available books at **www.hansebooks.com**

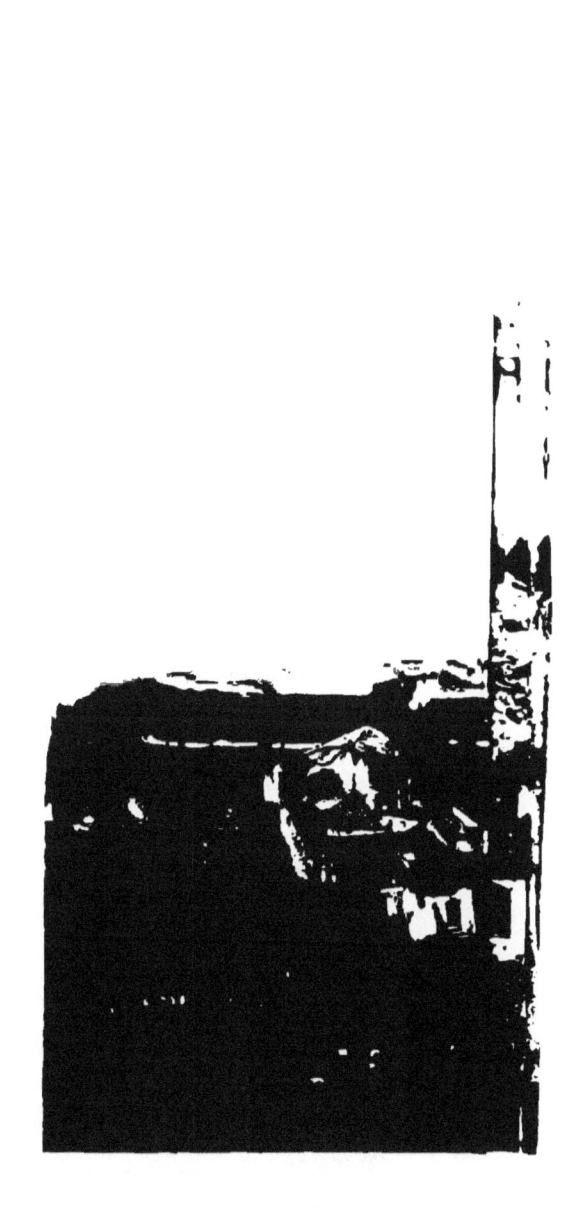

GREAT NEAPOLITAN EARTHQUAKE OF 1857.

THE FIRST PRINCIPLES

OF

OBSERVATIONAL SEISMOLOGY

AS DEVELOPED IN THE

REPORT TO THE ROYAL SOCIETY OF LONDON
OF THE EXPEDITION MADE BY COMMAND OF THE SOCIETY INTO
THE INTERIOR OF THE KINGDOM OF NAPLES,

TO INVESTIGATE THE CIRCUMSTANCES OF THE GREAT
EARTHQUAKE OF DECEMBER 1857.

BY

ROBERT MALLET, C.E., F.R.S., F.G.S., M.R.I.A.,
&c., &c.

"Non fingendum aut excogitandum sed inveniendum quid natura faciat aut ferat."

PUBLISHED BY THE AUTHORITY AND WITH THE AID OF THE
ROYAL SOCIETY OF LONDON.

IN TWO VOLUMES.—VOL. I.

CHAPMAN AND HALL, LONDON.
1862.
The Right of Translation is reserved.

DEDICATION.

To the Rev^D. T. R. Robinson, D.D., F.R.S., &c.,
Astron. Dir. Armagh.

My dear Sir,

Friendship, unbroken since my boyhood, and the many occasions upon which you have encouraged or assisted such attempts as I have made to advance knowledge, alone would induce me, with affectionate regard, to dignify these Volumes by connecting them with your own distinguished name; but to whom else could I so properly inscribe them, for to your early recognition of those truths which I had enunciated as the foundations of the science of Seismology, and to your prompt and weighty advocacy, of the importance to science, of seizing the opportunity of the great Neapolitan shock to apply those principles to nature, these Volumes probably owe their existence.

Ever, with much esteem,
I am yours,
ROBERT MALLET.

London, 1st October, 1862.

PREFACE.

SOME explanation seems desirable of how it is that the following Report to the Royal Society of London appears in form of the present volumes, and why not at an earlier period.

The earthquake of December, 1857, by almost the first notices that reached England, revealed itself as the third greatest in extent and severity of which there is any record as having occurred in Europe.

Impressed with its observational value to science, the Author at once addressed the following letter to the then President of the Royal Society :—

"DELVILLE, GLASNEVIN, CO. DUBLIN,
"28th *December*, 1857.

" MY LORD,

"The very recent occurrence of a great earthquake in the Neapolitan territory presents an opportunity of the highest interest and value for the advancement of this branch of Terrestrial Physics.

"Within the last ten years only Seismology has taken its place in cosmic science—and up to this time no earthquake has had its secondary or resultant phenomena—sought for, observed, and discussed by a competent investigator —by one conversant with the dynamic laws of the hidden forces we are called upon to ascertain by means of the more or less permanent traces they have left, as Phenomena, upon the shaken territory.

"Observed without such guiding light, or often passed by unnoticed and undiscovered for want of it—the facts hitherto recorded are in great part valueless—but with this guide such investigation is capable of results of high importance. Thus it was that Dolomieu's elaborate record of the effects of the great Calabrian Earthquake is of so much less value than it might have been.

"Earthquake observations are of two classes—those which *must* be made before and at the moment of shock (time and space measurements chiefly) and those which may be made at a recent period after it. To the latter belong those numerous and instructive facts treated of under the heads of Secondary and Accidental Phenomena in my First Report on the Facts of Earthquakes ('Reports Brit. Ass. 1850'), and also in the 'Admiralty Manual'—as well as many questions treated of under heads 15 'to 24' of the former. In those papers I have stated some of the methods of observation—of shattered buildings—altered water courses and springs—changed relations of level and position — localities of maximum and minimum disturbance—their relations to origin—to formation, &c., &c., and the inferences deducible. I need not, therefore, dilate upon them here.

"I have long looked for the occurrence of an opportunity so favourable for inquiry as that which has been just presented. It is one so rare, and in so peculiar and suggestive a region, that I venture to urge, through your Lordship, the Royal Society, that it should not be permitted to be lost to Science. To avoid this, however, the examination must be made with all possible *promptitude*, as every hour alters or removes the characters of the terrible inscription which we are to decipher, and renders circumstantial, local, and oral evidence less trustworthy.

"I respectfully offer, my Lord, if such be the will of the Royal Society, to proceed at once to Naples and the shaken regions, to collect, discuss, and report the facts. I am prepared to devote a month or five weeks to the inquiry, which, allowing ten days to the journey, to and from the city of Naples, I deem sufficient, if energetically and well employed.

"Were I a wealthy man I should proceed instantly, and on my own responsibility; but, although willing to give one-twelfth of my professional time and income for 1858 to it, private duties make it unsuitable that I should also be at the necessary expenditure for the journey and local inquiry.

"For this a sum of about One Hundred and Fifty Pounds would, I consider, be required, as the aid of local assistants, interpreters, with the peasantry, and the means of rapidly moving in remote and little frequented places (such as Basilicata), with other accessory charges, must be provided for. The best maps and all requisite instruments I am prepared with.

"In the humble but earnest confidence that I can in this do good service to Science, I submit to the Royal

Society whether it see fit to make such a grant, and to entrust the work to me; if so, I should be prepared to set out by the middle of next month.

"I have the honour to be, My Lord,

"Your very obedient Servant,

"ROBERT MALLET.

"*The Lord Wrottesley, President,*
   "*Royal Society, London.*"

The writer's views were promptly laid by the President before the Council of the Royal Society, where they were supported by the advocacy of Doctor Robinson, General Sabine, Sir Charles Lyell, Sir Roderick I. Murchison, and others. In result the Author was requested to proceed with the inquiry, and a grant made of the entire sum asked for. This amount proved, in the end, inadequate, as did also the period of time which it was proposed to devote to the investigation; so that the Author himself defrayed about two-fifths of the entire cost of the expedition, and found it necessary to devote to it more than double the time he intended beforehand. A further sum of Fifty Pounds toward procuring the Photographs, from which many of the Illustrations of these volumes have been reproduced, was voted by the Royal Society after the Author's departure, and the fact was communicated to him by telegraph.

A comparatively small proportion of the large collection of scientifically valuable Photographs made in the convulsed regions have been now reproduced, the limitation having been due to cost of production. The whole, however, are referred to by number in the text, those omitted with the addition (Coll. Roy. Soc.); and the originals may be consulted by Seismologists in the Library of the Royal Society, where the entire collection made, is now preserved along with the original sketches, maps, and manuscript.

Within a week of the Author's return to England in April, 1858, he addressed a preliminary Report to the President and Council of the Royal Society, the nature of which may be seen by the following extract from its introductory sentences :—

"On my return from the expedition of observation in the earthquake-shaken provinces of Naples, entrusted to me by the Royal Society, it becomes my duty to report briefly to your Lordship as President, and to the Council, where I have been, what I have done, and what of value to science, I may conceive myself to have accomplished; reserving scientific details and deductions to a separate and more systematic communication, which I hope ere long to have the honour of laying before the Royal Society, and which will embrace the results of my inquiries."

Two years nearly elapsed before the Author, interrupted frequently through necessity of professional avocation, and by other events, was enabled to complete the laborious work of reducing his observations, involving much tedious calculation, expanding and writing out his field notes, colligating his results, and finally bringing his Report into the form in which it was presented and read to the Royal Society on the 24th May, 1860, and which is substantially that in which it now appears in the following pages.

A brief abstract of the Report was published in the 'Proceedings Roy. Soc. Lon.,' vol. x. p. 486, &c., and the document itself, with its accompanying maps, diagrams, sketches, and photographs, ordered to be referred for publication.

As to the mode of this, some difficulty arose. The Report, although in a great degree dealing with dynamical and other rigid questions, is partly an inductive argument; and hence, being necessarily of the nature of a *pièce justicatif*, requires the sufficient statement of a number of directly employed facts; besides and related to which were, many observations made *en passant*, upon the physical features, geology, and other collateral subjects, which, referring to a country so little explored as the interior of the Neapolitan kingdom, it was not desirable to suppress.

Its bulk thus became such that if published by the Royal Society in what might seem its natural place, it would have occupied an entire volume of the 'Philosophical Transactions,' to the exclusion, for the time, of all other papers, however valuable. Some suggestions were made to divide the Report, publishing in the 'Philosophical Transactions' only its rigid mechanical portions in result, leaving the remainder for publication there in future years; but to this the Author felt much objection.

Finally, the Council of the Royal Society, in the exercise of a wise discretion, and in a very liberal spirit, decided to devote a sum of three hundred pounds towards the cost of illustration, and proposed to the Author that he should publish, *in extenso*, his Report as a distinct and independent work.

In pursuance of this, the Author arranged with Messrs. Chapman and Hall for publication in the form in which the work now appears, the Publishers having undertaken to defray the entire expense over and above the grant made by the Royal Society. Liberal as the amount of that was, it was not nearly sufficient for its object, owing to the great expenditure involved in the production of volumes illustrated as these are. The Author can but hope that the spirit with which the Publishers have thus shown themselves ready substantially to aid in promoting science, may not be to them profitless.

It was midsummer last, before all these preliminaries admitted of the work being placed in the hands of the printer, and the illustrations in those of Mr. V. Brooks for lithographing. Practically, the whole was ready for publication early this year. For trade reasons the Publishers requested that it might be postponed to the present time, to which the Author, though conscious of the long delay that circumstances had already imposed, readily assented.

These volumes, as they now appear, will probably be perused by two distinct classes. To the first, the really scientific reader, the Author will venture, though perhaps at the risk of some undeserved suspicion of egotism or vanity, to commend the subject and the method which they evolve, as pregnant with the power of future knowledge, of the cosmical conditions of the interior of our own and of other planets; that will be hereafter recognized as having first shown the way to any true intelligence from the viewless and unmeasured miles of matter beneath our feet; and that will ere long give us up the key to the hitherto undeciphered enigma of vulcanicity.

To the general reader, earthquake narratives have long shared in some degree, the charm that belongs to tales of shipwreck, of battle, or wild adventure, and "perilous hair-breadth 'scapes" amidst natural phenomena the most tremendous: something of this he here will find; and though sobered to a reality not always found in earthquake stories, the events by which such multitudes perished, in which so many cities were overthrown, will be found by him who shall have even generally understood the principles here unfolded, to yield more intelligent pleasure, than the exaggerated and often fabulous phantasmagoria, of the older earthquake narratives, in the maze of which he wandered without any rational clue.

He will here trace with interest the successive steps by which finally the depth of the focus, whence the impulse that produced the earthquake has been for the first time ascertained, and measured in miles and yards with the certainty that belongs to an ordinary geodetic operation.

It is of the nature of all science, to be but the portal to greater and higher truth beyond. Such is peculiarly the case with Seismology. The exact knowledge of earthquakes, of their distribution in time and space, of their movements, results, and proximate causes, however interesting in themselves, are yet but means to an end.

As palæontology—itself dependent upon natural history—lithology, and many other cognate knowledges, are but instruments of geology, so is seismology chiefly to be viewed and valued, as the instrument by which a knowledge of the deep interior of our planet will be attained; the only instrument yet discovered to this end, yet one possessed of vastly greater power and directness of aim, than any of those that physical geology has previously called to its aid.

Though the youngest branch of cosmical science, it is to be regretted, that it has not been already better understood, and more applied by observational geologists, many of whom, had they mastered even its rudiments, might ere now have come laden with fruit from various regions.

Physical geology, much as it owes to the labours, of the topographical and field geologist, to the patient observer and comparer of nature's superficial phenomena, can no longer rest satisfied with such modes of investigation alone. The time has more than come, when it must devise new methods and new problems, and appropriate to their solution, all the aids that theoretical mechanics, including those of undulations, physics, and chemistry can afford; and the geologist of the coming time, who shrinks from mastering these, though he may continue a labourer, shall be no "light bearer" in the rising palace of cosmical truth. More or less, it is the Author's hope that these volumes may be ancillary to promoting and giving direction to such a result.

London, October, 1862.

LIST OF ILLUSTRATIONS

USED OR REFERRED TO

IN THE FIRST VOLUME.

Note.—In the following LIST OF ILLUSTRATIONS, and in the text, wherever the reference is made in the words, "*Collection of the Royal Society*," or "*Coll. Roy. Soc.*," it is to be understood that such Illustrations have been necessarily omitted from this work, in order to limit the expense of reproducing so great a number of Photographs or Sketches, and that the originals of all such as are so omitted are to be found in the possession of the *Royal Society of London*.

*** For List of Maps *see* Illustrations to Vol. II.

City of Polla after the Earthquake.—*See* No. 161 . . . *Frontispiece*.

Number.			Page.
1. Woodcut		Dalziel Bros.	16
2. „		„	16
3. „		„	18
4. „		„	18
5. „		„	18
6. „		„	20
7. „		„	20
8. „		„	21
9. „		„	21
10. Collection of the Royal Society			25
11. „ „			26
12. „ „			27
13. Castelluccio		*V. Brooks, to face*	29
14. Padula		„	30
14 *bis*. Woodcut		Dalziel Bros.	33

VOL. I.

LIST OF ILLUSTRATIONS.

Number.			Page.
15.	Collection of the Royal Society		35
16.	„	„	35
17.	„	„	35
18.	„	„	35
19.	„	„	35
20.	„	„	35
21.	Woodcut	Dalziel Brs.	38
22.	„	„	38
23.	„	„	38
24.	„	„	38
25.	Church at Pertosa, looking North-West	V. Brooks, to face	42
26.	Woodcut	Dalziel Brs.	45
27.	„	„	46
28.	„	„	49
29.	„	„	51
30.	Collection of the Royal Society		51
31.	Woodcut	Dalziel Brs.	53
32.	„	„	53
33.	„	„	53
34.	„	„	53
35.	„	„	60
36.	„	„	59
37.	„	„	59
38.	„	„	59
39.	„	„	60
40.	„	„	60
41.	„	„	63
42.	„	„	63
43.	„	„	65
44.	„	„	72
45.	„	„	67
46.	„	„	68
47.	„	„	73
48.	„	„	66
48 bis	„	„	69
49.	Auletta	V. Brooks, to face	73
50.	The Cathedral, Paterno	„	74
51.	Collection of the Royal Society		74
52.	Woodcut	Dalziel Brs.	76
53.	„	„	76
54.	„	„	77
55.	„	„	77
56.	„	„	77

LIST OF ILLUSTRATIONS.

Number.			Page.
57. Woodcut		Dalziel Bros.	77
58. „		„	77
59. „		„	86
60. „		„	87
61. Cathedral of Marsico Nuovo, North side.—See page 42, vol. i.		V. Brooks.	
62. Santa Dominica, Montemurro, looking South, from the Palazzo Fino		„	89
63. At Polla		„	93
64. Collection of the Royal Society			93
65. Woodcut		Dalziel Bros.	97
66. Polla, Strada Rorco		V. Brooks, to face	99
67. Collection of the Royal Society			99
68. „ „			99
69. Woodcut		Dalziel Bros.	101
70. „		„	103
71. „		„	103
72. „		„	103
73. „		„	103
74. „		„	103
75. „		„	103
76. „		„	105
77. „		„	105
78. „		„	109
79. „		„	109
80. Church at Picerno.—See page 89, vol. i.		V. Brooks.	
81. Interior of the Cathedral at Tito, looking north-west.—See page 99, vol. i.		„	
82. Collection of the Royal Society			114
83. Church of the Madonna di Lorretto, Polla.—See No. 168, page 296, vol. i.		V. Brooks	
84. Collection of the Royal Society			115
85. „ „			
86. Woodcut		Dalziel Bros.	116
87. „		„	117
88. „		„	117
89. „		„	118
90. Collection of the Royal Society			118
91. Woodcut		Dalziel Bros.	119
92. „		„	120
93. „		„	125
94. „		„	126
95. „		„	127

LIST OF ILLUSTRATIONS.

Number.			Page.
96.	Woodcut	Dalziel Brs.	129
97.	,,	,,	130
98.	,,	,,	131
99.	,,	,,	133
100.	,,	,,	134
101.	,,	,,	135
102.	,,	,,	135
103.	,,	,,	140
104.	,,	,,	145
105.	,,	,,	155
106.	,,	,,	155
107.	,,	,,	158
108.	The Val di Diano, Town of Diano opposite	V. Brooks, to face	165
109.	Collection of the Royal Society		165
110.	,, ,,		169
111.	Woodcut	Dalziel Brs.	204
112.	,,	,,	204
113.	,,	,,	204
114.	,,	,,	208
115.	,,	,,	214
116.	,,	,,	218
117.	,,	,,	218
118.	,,	,,	225
119.			
120.	Vietri, near the New Road	V. Brooks, to face	232
121.	Woodcut	Dalziel Brs.	234
122.	,,	,,	234
123.	From the Plain of Pæstum.—See p. 272, vol. i.	V. Brooks.	
124.	,, ,, ,,	,,	
124 bis.	East Flank, Eboli	,,	240
125.	West Flank, Eboli	,,	240
126.	Collection of the Royal Society		242
127.	Woodcut	Dalziel Brs.	241
128.	,,	,,	246
129.	,,	,,	251
130.	,,	,,	251
131.	,,	,,	253
132.	Woodcut	Dalziel Brs.	256
132 bis.	Auletta, showing the directions of the landslip and long fissures in the soil—Eye sketch	V. Brooks, to face	257
133.	Collection of the Royal Society		258
134.	,, ,,		260

LIST OF ILLUSTRATIONS.   xxi

Number.		Page.
135. Auletta	*V. Brooks, to face*	260
136. At Auletta	„	260
137. Woodcut	*Dalziel Bro.*	262
137 bis. Villa Carusso, near Auletta	*V. Brooks, to face*	267
138. The Porte Cochere, on the Military Road, Villa Carusso, near Auletta	„	270
139. Woodcut	*Dalziel Bro.*	268
140. „	„	269
141. „	„	269
142. „	„	271
143. Collection of the Royal Society		273
144. „ „		273
145. The Flanks of Monte Alburno and Castelluccio, near Auletta	*V. Brooks, to face*	272
146. Collection of the Royal Society		274
147. „ „		274
148. „ „		274
149. „ „		274
150. „ „		274
151. Pertosa	*V. Brooks, to face*	274
152. Woodcut	*Dalziel Bro.*	275
153. „	„	279
154. Section of the Valley at Pertosa	*V. Brooks, to face*	282
155. Campostrina, Gorge of the Calore. Great Fall of Rock	„	285
156. Campostrina	„	286
157. Woodcut	*Dalziel Bro.*	287
158. Fissures on the Road near Polla	*V. Brooks, to face*	288
159. Tenementa della Madonna, Campostrina.—See page 274, vol. i.	„	
160. Collection of the Royal Society		289
161. (See Frontispiece)	*V. Brooks.*	
162. Collection of the Royal Society		294
163. „ „		294
164. „ „		294
165. „ „		294
166. „ „		294
167. Monastery of St. Claire, Polla, looking South-west	*V. Brooks, to face*	295
168. Church of the Madonna di Loretto, Polla	„	296
169. Collection of the Royal Society		297
170. Polla.—See page 288, vol. i.	*V. Brooks.*	
171. Collection of the Royal Society		298

LIST OF ILLUSTRATIONS.

Number.		Page.
172. Collection of the Royal Society		298
173. Small House by the River, Polla, looking westward	*V. Brooks, to face*	299
174. House on the Bank of the River, Polla, looking eastward	,,	299
175. Ground Plan of the Palazzo Palmieri	,,	302
175 bis. Palazzo Palmieri at Polla	,,	302
176. Collection of the Royal Society		304
177. Interior Façade, Palazzo Palmieri, Polla, looking westward		303
178. Interior Court, Palazzo Palmieri, Polla.—*See* page 286, vol. i.	*V. Brooks.*	
179. Interior Court, Palazzo Palmieri, Polla	,,	303
180. Collection of the Royal Society		304
181. Woodcut	*Dalziel Bro.*	306
182. ,,	,,	306
183. Camine, Palazzo Palmieri, Polla, looking south.—*See* page 10, vol. ii.	*V. Brooks.*	
184. Collection of the Royal Society		314
185. ,, ,,		313
186. ,, ,,		314
187. Atena.—*See* page 286, vol. i.	*V. Brooks.*	
188½ Atena	,,	326
189. Collection of the Royal Society		324
190. A Street in Atena.—*See* page 326, vol. i.	*V. Brooks.*	
191. Collection of the Royal Society		327
192. ,, ,,		327
193. Woodcut	*Dalziel Bro.*	328
194. ,,	,,	330
195. ,,	,,	330
196. ,,	,,	333
197. Collection of the Royal Society		334
198. ,, ,,		335
199. ,, ,,		335
200. Woodcut	*Dalziel Bro.*	338
201. ,,	,,	339
202. ,,	,,	339
203. ,,	*Dalziel Bro.*	341
204. ,,	,,	342
205. Collection of the Royal Society		341
206. Woodcut—Church of La Sala	*Dalziel Bro.*	343
207. Collection of the Royal Society		345
208. Woodcut	*Dalziel Bro.*	347

LIST OF ILLUSTRATIONS. xxiii

Number.			Page.
209. Collection of the Royal Society			349
210. Valley to the east of La Sala—Val. di Diano		V. Brooks, to face	352
211. Rock Aiguille, near Padula		,,	353
212. ,, fractured from its base		,,	353
213. Collection of the Royal Society			355
214. ,, ,,			355
215. ,, ,,			356
216. Woodcut		Dalziel Brs.	356
217. Overthrown Column, Palazzo Romani, Padula. —See page 34, vol. ii.		V. Brooks.	
218.			
219.			
220. Collection of the Royal Society			362
221. Woodcut		Dalziel Brs.	359
222. ,,		,,	363
223. ,,		,,	366
224. ,,		,,	367
225. Monument of St. Bernard, at the Certosa, Padula		V. Brooks, to face	370
226. Collection of the Royal Society			371
227. ,, ,,			371
227½. ,, ,,			371
228. ,, ,,			372
229. ,, ,,			373
230. ,, ,,			373
231. ,, ,,			373
232. ,, ,,			373
233. ,, ,,			373
234. Grand Certosa, Padula. Interior of the Gallery of the Grand Court.—See page 13, vol. ii.		V. Brooks.	
235. Woodcut		Dalziel Brs.	378
236. ,,		,,	379
237. ,,		,,	380
238. Diagram, The Campanile, Cistercian Monastery, Padula		V. Brooks, to face	370
239. Woodcut		Dalziel Brs.	372 & 392
240. Diagram, Cistercian Monastery, Padula		V. Brooks, to face	370
241. Diagram, Section from the Valley of the Calore to that of the Agri		,,	373
242. Diagram, Section from Spinosa to Viggiano, across the Agri		,,	403
243. Collection of the Royal Society			411
244. Woodcut		Dalziel Brs.	414
245. ,,		,,	414

LIST OF ILLUSTRATIONS.

Number.			Page
246. Collection of the Royal Society	.	.	414
247. „ „	.	.	414
248. Sarconi, Remains of the Church	.	*V. Brooks, to face*	415
249. Collection of the Royal Society	.	.	416
250. Woodcut	.	*Dalziel Brs.*	415
251. Saponara, with the remains of the Castello Cilliberti, looking Northward	.	*V. Brooks, to face*	419
252. Saponara, Remains of the Church.—*See* page 415, vol. i.	.	„	
253. Mound of Rubbish where Saponara had been, looking Southward	.	„	419
254. Collection of the Royal Society	.	.	420
255. Saponara	.	*V. Brooks, to face*	424
256. „	.	„	424
257. Woodcut	.	*Dalziel Brs.*	426
258. Beds of the Agri and Moglia, Denudation of the Breccia.—*See* page 424, vol. i.	.	*V. Brooks.*	

INTRODUCTORY.

On the 16th December, 1857, an earthquake of great violence visited several of the southern provinces of the Neapolitan kingdom. Accounts of the formidable extent of the disaster, accompanied by a few imperfect details as to some of its physical phenomena, began to arrive in England, through correspondence and the public press, about the 24th December. The occasion appeared to the author to present an opportunity of observation of the highest value for the advancement of our knowledge of earthquakes, considered as a branch of cosmical science.

On the 28th of the same month he accordingly addressed a letter to Lord Wrottesley, President of the Royal Society, suggesting the importance to science of sending a competent observer, without loss of time, to the convulsed region, and offering, with the approval and assistance of the Society, to undertake the duty. The suggestion was promptly laid before the Council by the President, met its approval, and on the 21st January, 1858, the author received the authority of the Royal Society to proceed.

From the 21st to the 26th January was occupied in obtaining letters and recommendations from the Council and officers of the Royal Society, the British Minister for Foreign Affairs, and some noble or eminent scientific persons in London, to the Government of the kingdom of Naples, its well-known jealousy causing much doubt whether its permission might be obtained for the author to travel into the earthquake districts.

On the 27th January the author left London, stopping a day at Paris, for the purpose of conferring with and receiving any suggestions that those there engaged in geological research might have to offer as to his intended labours; and another at Dijon, in conference with Professor Perrey, the distinguished author of many works on seismology, with a like object. On the 5th February, 1858, he arrived at Naples, where he was detained until the 10th, awaiting the tedious decision of the Neapolitan Government as to whether or not it would permit his journey into the interior. This permission was at length granted by telegraph from the king, then at Gaeta; and, accompanied ultimately by letters of authority to the Intendenti, Judici, Syndici, and Gendarmerie of all the provinces proposed being traversed, enjoining them to give safe conduct and all possible assistance to the author and his objects.

Before leaving England he had been favoured by Cardinal Wiseman with an encyclical letter, commending the object of his mission to the good offices of the clergy of all denominations in the interior. This valuable letter he was enabled, after an interview with the Cardinal (Sisto) Archbishop of Naples, to get approved and countersigned by that prelate; and to that document he owes the removal of many difficulties that might otherwise have seriously impeded his progress and means of observation.

The interval caused by the delay of the Neapolitan Government before granting its authorization to proceed, was occupied by the author, partly in providing, with the kind assistance of some British residents at Naples, a suitable and trustworthy staff of persons to accompany him, and amongst these an interpreter who could converse readily in the provincial dialects of the Capitanatas, Basilicata, Bari, and Calabria, and in providing, with the requisite forethought, all the camp and cooking equipage, blankets, food, medicines, and the means of their convenient transport upon mules through a rough and mountainous country, of which large tracts were expected, and were in the end actually found to be, destitute (as a consequence of the earthquake) of either shelter or provisions.

The remainder of the time was employed in making and recording observations and collecting information as to the phenomena of the earthquake, as it was felt in and around the city of Naples, and its immediate neighbourhood, and in seeking for evidence along the shores of the bay, from Pozzuoli to Castellamare, of recent change of level of the land, traceable to the shock. Arrangements were also provisionally made with an eminent French photographer at Naples for his following the probable track of the author, and obtaining a series of photographic views of such scenes and objects as might illustrate the results of the expedition; and the author wrote home requesting the sanction of the Council of the Royal Society to his obtaining the photographs which now are referred to and inwoven with this Report.

Before leaving the city of Naples, Professors Palmieri and Scacchi were consulted by the author, for any suggestions that they might deem desirable to make as to his examination, such as might occur to them from their prolonged familiarity with the seismic conditions of their country, and to be expected from their scientific eminence. He has to thank both for their cordial reception, and especially to record his obligations to Signor Guiscuardi, whose accurate and extensive scientific information and local knowledge were of great service to him in making his arrangements with the Government officials, and whose zealous kindness the author retains in lively remembrance.

The following Report is divided into three principal parts. In the first, the questions proposed for observation or solution are generally stated, and the methods of observation pursued in these earthquake regions to some extent indicated, with some remarks upon the physical and other characteristics of the country itself, in relation to seismical effects. The second part embraces the narration of the observations actually made and information obtained, with some primary discussion of facts in a few instances. And in the third, the facts, so far as they admit of it, are classified and put together, and such conclusions or deductions drawn from them as they appear to the writer to warrant.

In these the author will, for convenience and simplicity, write in the first person.

PART I.

CHAPTER I.

OF THE QUESTIONS FOR INQUIRY, AND METHODS OF OBSERVATION, OF EARTHQUAKE PHENOMENA.

WERE Seismology an older and more mature branch of science than it is, it would be impertinent to enter at any length into the means and methods by which it is to be pursued, in the observation of earthquake phenomena. Dating, however, for anything approaching to scientific guidance or precision, not more than twelve years back,* it is the more necessary to make generally intelligible the methods of observation which can be pursued in a seismic region *after* the occurrence of the shock, in order that the evidence upon which conclusions may be drawn as to the direction, velocity, amount of movement, &c., of the latter may be accepted with their just weight; and the rather, because as yet it is not to all persons quite self-evident how *any* information whatever can be had or conclusions drawn as to a phenomenon so perfectly transient and momentary as an earthquake shock, by examination, at a considerable time after its occurrence, of the region over which it has passed.†

* This was written in 1858.
† With the exception of the author's 'Instructions' in the 'Admiralty Manual,' nothing whatever has been written bearing upon

An earthquake, like every other operation of natural forces, must be investigated by means of its phenomena or effects. Some of these are transient and momentary, and leave no trace after the shock, and such must either be observed at the time, or had from testimony. But others are more or less permanent, and, from the terrible handwriting of overturned towns and buildings, may be deciphered, more or less clearly, the conditions under which the forces that overthrew them acted, the velocity with which the ground beneath was moved, the extent of its oscillations, and ultimately the point, can be found in position and depth beneath the earth's surface, from which the original blow was delivered, which, propagated through the elastic materials of the mass above and around, constituted the shock.

Again, certain effects, such as landslips, fissures, alterations of water-courses, &c., are produced of greater or less permanency affecting the natural features of the shaken country.

The observation of each of these classes of effects bears reference to two distinct orders of seismic inquiry.

By the first, we seek to obtain information as to the depth beneath the surface of our earth at which those

---

such methods of observation. Mr. Hopkins, in his mathematical *résumé* of the laws of elasticity, as bearing upon seismology, communicated to the British Association (17th Report, Oxford, 1847), incidentally points out the geometric conditions by which, if the emergence of a shock were known, the depth of the origin might be ascertained. The procedure generally had been previously suggested by the author in his original memoir on 'Dynamics of Earthquakes.' Trans. Roy. Irish Acad., vol. xxi., Part 1. The methods employed in this work are altogether distinct from that noticed by Mr. Hopkins.

forces (whether volcanic or otherwise) are in action, whose throbbings are made known to us by the earthquake, and thus to make one great and reliable step towards a knowledge of the nature of these forces themselves; and this is the great and hopeful aspect in which seismology must be viewed and chiefly valued. It affords, if not the only, certainly, in the existing state of knowledge, the best means by which we can entertain a well-founded expectation of ultimately obtaining clear and certain ideas as to the material and state of the internal mass of our planet, and comprehending the true nature and relations of volcanic energy.

By the second order of inquiry we seek to determine the modifying and moulding power of earthquake upon the surface of our world as we now find it; to trace its effects and estimate their power and extent upon man's habitation and upon himself. The first order of inquiry must be pursued by methods, chiefly mechanical, physical, or mathematical. The second by these, combined with the observational tact and largeness of a disciplined imagination and eye that are amongst the accomplishments of the physical field-geologist. Thus finally uniting our knowledge derived from both directions, ultimately to form a clear conception of what is the function of the earthquake in the Cosmos, and to recognize the connection, fitness, order, and beauty, even of the volcano and the earthquake, as parts of the machinery of a wondrous and perfect creation. Like every aspect of nature, that we obtain with the more enlarged and undimmed eye of truth, it will prove to us that even here the great Author of all, is a God of order, not of confusion.

These, then, were the questions that I set before me. To endeavour to find the position, superficially and in depth of the centre of impulse of the shocks of the 16th December 1857, and to observe and discuss the effects of the earthquake, actual and prospective, upon the face of the country, in relation to all its physical conditions.

The method of investigation which I purposed to adopt is based upon the very obvious truth, that the disturbances and dislocations of various solid objects by the shock of earthquake, if carefully observed with reference to their directions and extent of disturbance, and to the mechanical conditions in play, must afford the means of tracing back from these effects, the directions, velocities, and other circumstances of the movements or forces that caused them. This mode of examination, strange to say, appears to be perfectly new, and to have escaped the attention of all previous examiners of earthquake-shaken districts, as well as of all writers upon the subject. Thus the government reporters (for example) upon the great Calabrian earthquake of 1783, or those more recently (Palmieri and Scacchi) upon that of Basilicata in 1851, seem to have been perfectly unconscious that in the fractured walls and overthrown objects scattered in all directions beneath their eyes, they had the most precious data for determining the velocities and directions of the shocks that produced them.

The idea of applying number and measure to these never seems to have occurred to them. They merely describe the particulars in a loose and general way, and only occasionally as curious, or remarkable, or inexplicable exemplars of the power of the disturbance. Hence they failed to draw a single conclusion of certainty or scientific value as to the place whence the shock emanated, how deep this was under the earth, or in what direction it emerged from beneath it.

As this method of seismic observation then, is novel, as I trust to show that it has proved fruitful in result, in this its very first application to nature; and as, when fully developed, it will be found a real "Organon," a powerful machine for future discovery, I make no scruple in treating at considerable length of its methods, in the hope that they may become understood, diffused, and applied by others. This method, I believe, will be hereafter recognized as one of the most fruitful applications yet made of mathematics to physical geology.

Before leaving England for Naples, I communicated my views as to this method of investigation to my friend the Rev. Samuel Haughton, F.R.S., Professor of Geology, Trinity College, Dublin, and requested him to arrange for me a series of workable equations that should embrace most of the conditions as to direction and velocity of fractured or overthrown bodies that I expected to meet. With the utmost readiness, he applied his adroit and eminently practical mathematical powers to the task, and from him I received the equations given at pp. 125 *et seq.*—I. to XLV.—which formed some of the most valued working tools of my deductions.

CHAPTER II.

ELASTIC WAVE OF SHOCK.

THE elastic or earth wave of shock, may reach a given point upon the surface, with any angle of emergence (the angle contained by the horizontal plane with the wave-path at the point of emergence), or in any azimuth. The path of the wave is a right line joining that point with the centre of impulse (or focus), the wave being assumed propagated thence in all directions outwards in spherical shells. This is strictly true only in an homogeneous elastic solid.

Every point in a coseismal line (that in which such a wave shell simultaneously reaches the earth's surface) at the moment of shock describes a closed curve in space, returning to (or almost exactly to) the point from which it started into motion. The curve is one of double curvature, the vibration taking place nearly simultaneously, in three rectangular unequal axes. For the purposes of our inquiry we may neglect the transversal vibration, and may consider the closed curve of normal vibration as confined to vertical planes passing through the centre of impulse. In fact, the movement of the wave particles may be assumed as confined to right lines, *indirectum* with the path of the wave, whose length is equal to the amplitude of the wave or to one half of its complete vibration.

As the coseismal curve (or crest of a wave of shock) enlarges its area, travelling outwards in all directions from the seismic vertical—that is, from the vertical line passing through the earth's surface (and centre) and the focus—every point in and upon the surface, in succession moves once forward and back in the direction of the wave-path, and to the extent of its amplitude at that point, or in two components, vertical and horizontal, that shall give such direction.

We are not concerned here to consider the extent or the laws according to which the range of wave movement of each material particle, either in amplitude or attitude, or the velocity of its particles, vary with the distance from the focus.

The angle of emergence for any given depth of focus diminishes as a function of the distance of any given point of the surface from the seismic vertical; and I have shown elsewhere that the power of the shock to overthrow objects is, *cæteris paribus*, a maximum at a determinate distance all round the seismic vertical; and this distance would be equal all round, were the earth homogeneous, and the focus or centre of impulse confined to a mathematical point. The centre of impulse in nature, however, occupies determinate, and often large dimensions. For this reason as well as from non-homogeneity amongst others, neither the meizoseismic curve (or that of maximum overthrow) nor the isoseismic curves (or those of equal overthrow) are found to be circles or even perfectly regular closed curves, nor concentric.

For the same reason the observed angles of emergence will be found to vary from those that would be due to a focus of evanescent magnitude.

The distinction must be clearly borne in mind between the velocity of transit of the wave—that with which the advancing form or seismal curve is transferred from point to point of the surface, and that of the earth particles moving within the limits of amplitude of each vibration. The former velocity is very great, nearly half as rapid as that of a cannon shot, and depends chiefly upon the elastic modulus of the earth's formations through which the wave transit is made; but the latter, as now measured for the first time, is very small indeed, often not greater than that which a body acquires by falling from a height of two or three feet.

It is, however, to the rapidity of transit velocity, or, which is the same thing, to the great rapidity with which the proper velocity of vibration of the wave passes, from 0 to its maximum velocity, on reaching any material object, that the formidable dislocating effects of the very moderate maximum velocity of vibration are due. We need not, however, here extend these preliminary remarks.

The evidences fitted for observation after the shock, by which the conditions of earthquake motion are discoverable, may be divided into two great classes:—

1st. Fractures or dislocations (chiefly in the masonry of buildings) which afford two principal sources and sorts of information.

> *a*. Information from the observed *directions of fractures or fissures*, by which the *wave-path*,

and frequently the *angle of emergence* may be immediately inferred.

    b. Information from the preceding united with known conditions as to the strength of materials to resist *fracture*, by which the *velocity* of the fracturing impulse may be calculated.

2nd. The overthrow, or the projection or both, of bodies, large or small, simple or complex. From these we are enabled to infer—

    c. By direct observation the direction in *azimuth* of the wave-path.

    d. By measurements of the horizontal and vertical distances of overthrow or of projection, to infer the *velocity* of projection and *angle of emergence*—both, or either.

Fractures or dislocations present themselves always in directions more or less *transverse* to the wave-path. Overthrow or projection, on the contrary, always takes place in the line of the wave-path, or in the vertical plane passing through it; but the *direction* of fall or of projection may be reverse (or in the contrary direction) to that of the wave transit, or it may be in the same direction with it.

At the moment of the arrival of the earth wave at any object upon the surface, whose dimensions are less than the amplitude—an obelisk or pillar or single wall for example—motion is suddenly communicated to the body; the velocity of the vibrating particles rapidly increases from zero to its maximum velocity, and returns to zero, as it completes its *first semiphase* or half-vibration, *the direction of movement in which is in the same sense as that of the wave transit.*

With nearly the same rapidity the velocity increases in the opposite direction from zero to the maximum, and back to zero again. The wave has then passed the given point, its whole phase or entire vibration has been completed, and it has produced its effects. The movement applied, is opposed by the inertia of the body moved, whose motions and final displacement depend upon the direction of the wave-path with regard to the centre of the body, its form, and the position of its base, or points of adherence or of support, and to the maximum velocity of the wave's proper motion. The applied velocity acts at the centre of gravity and in the direction of the wave-path, and the body, if free, *apparently* moves in the *opposite direction* to *the wave* in its first semiphase in consequence of its inertia. The force of displacement, with a given maximum velocity of vibration is therefore always proportionate to M, the mass, so that a heavy body, in the same shape and conditions is as easily upset as a light one.

If the body be not free, if the line of wave transit passing through its centre of gravity pass within the base or through any other support, it does not move in the first semiphase of the wave; but if it be free in the opposite direction, it will be displaced in the second semiphase of the wave; but as the wave movement is now reverse to that of its transit, the inertia of the body acting still contrary to the applied velocity, now impels the body in the *same direction* as the *wave transit*.

In either case, and in either semiphase of the wave, the movement impressed, may be one of mere overthrow or upsetting, or it may be one of actual projection, or of both combined, depending upon the special conditions of the body and its supports, &c.

Where the body is projected from a base or support with which it has had friction or adherence, and that the line of wave transit through its centre of gravity does not also pass through the centre of adherence (that is, the point of the base, or between it and supports, in which all the resisting forces, of adherence, &c., may be supposed concentrated), then, besides projection, a movement round a centre of spontaneous rotation within the body will also be impressed. Where this is due to adherence at the base, the rotation is generally in a vertical plane, and does not seriously disturb the plane of projection from that of the wave-path, *i.e.*, of a vertical plane passing through the seismic focus and the body displaced; but when also due to lateral adherence, or other still more complex conditions, the body is flung forward and whirls round on inclined axes, and finally comes to rest in some position quite abnormal to its original status, giving rise occasionally to complex phenomena from which nothing can be inferred. Where the body is large, such as a house or church of masonry, or even a single wall exposed to shock in the plane of its length, overthrow may be impossible with given dimensions and given angle of emergence of the wave; but in such case dislocation or fissuring occurs, and the severed parts may or may not be overthrown, dependent upon the amount of the applied velocity consumed in producing fracture only.

There may be no displacement whatever of loose objects, nor any dislocation of large masses, such as churches, &c., though exposed to violent shock, if its emergence be quite or very nearly vertical, and that the maximum velocity of the wave does not exceed—

$$V = \sqrt{2gH}.$$

H being equal λ, the amplitude of the wave, the masses being in such case rapidly lifted up and let fall again without the withdrawal of the support of the base.

And generally, single objects situated upon the surface of the earth, in firm and rigid connection with it, or so circumstanced that the line of wave transit through the centre of gravity passes through the surfaces of repose and of attachment, move with the earth itself, and are seldom disturbed as to their former position. Thus also, flexible objects, rooted trees, flag staffs, telegraph posts, and the like, are bent by the transverse forces impressed, but return to their positions, leaving only perhaps traces in the earth disturbed at their bases, of the direction of movement.

A few examples may clear this part of the subject. In Fig. 1 let there be a large stone ball, adherent to a narrow base on top of its pedestal, which is fast in the

Fig. 1.   Fig. 2.

ground, and exposed to a shock, the direction of wave motion of which, in the first semiphase, is from *a* to *b*, and with velocity sufficient to dislodge it. The ball, urged by inertia, and free to fall in any direction, will be projected in *b* to *a*, contrary to the wave, and describing a trajectory in the plane of the wave-path, will fall to *c*, and if its velocity in the horizontal axis be not wholly destroyed (as by falling on soft ground), it may roll, and so cease to give any indication of subsequent value. We are not now concerned with what is the trajectory, or with its modifying conditions. It is obvious that these might be such that the ball shall drop, nearly plumb to the ground, but always in a direction contrary to that of the wave transit.

If a similar ball however, and exposed to a like shock, have a support—such as a wall, for example, at the side *a* (Fig. 2), which is not overthrown by the shock, but carried along with the wave in the forward movement of its first semiphase—then the ball, although pressed by inertia against the adjacent face of the wall, is prevented falling in that direction. It also, therefore, has impressed upon it, the maximum velocity of the wave in its first semiphase. When, therefore, the wave itself arrives at its maximum velocity in the contrary direction—viz., in its second semiphase—the ball, by its inertia of motion impressed in the previous semiphase, is then thrown in the *same* direction as the wave transit *a* to *b*, and projected to the ground at *d*, as before.

It might happen that the wall might be so related in dimensions, &c., to the velocity and direction of the wave, that it should remain standing long enough to produce the effects described upon the ball, but should immediately afterwards begin to fall, fracturing and turning over upon the point *f* in a direction contrary to that of the wave transit; and in such a case there might remain no evidence to show, from either the ball or the wall, in which direction the wave transit was made, whether from *a* to *b* or the contrary. We might ascertain the path of the wave, or rather the azimuth in which it lay, but no more.

If at *two* distant localities we can obtain even that much information, we can assign the place of the seismic vertical. For if (Fig. 3) by one shock the ball B be projected in the wave-path *a'* to *b'*, in either direction, suppose contrary to the wave transit, and also the ball A at a distant place not in the same right line, also projected, say either as the

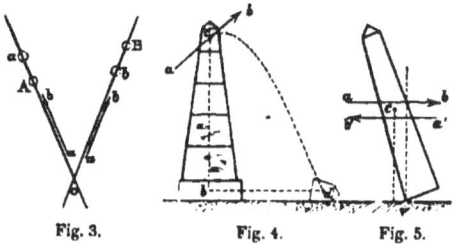

Fig. 3.   Fig. 4.   Fig. 5.

former or in the same direction as the wave transit *a* to *b*, then we obtain two azimuths, which can have but one point of intersection in o, which is that through which the seismic vertical passes.

Where the body projected (Fig. 4) is circumstanced so as to retain the position in which it alighted upon the ground, so that we can measure the vertical and horizontal axes *b c*, *b d*, then knowing the maximum velocity of the wave, and which is equal that of projection, we can find the angle of emergence in the plane of projection whose azimuth is observed, and *vice versâ*.

And if we have two closely adjacent objects projected in the same locality, and the above conditions observed, we can calculate both the angle of emergence and the velocity.

It will most generally happen that a regular solid (such as an obelisk, &c.) will fall prostrate whenever the maximum velocity of the wave is such as to produce in it oscillation sufficient to destroy statical equilibrium; and as the arc of oscillation due to a given velocity may be assigned, if we know the angle of emergence of the wave, so as to arrive at its horizontal component of velocity, we can always assign an *inferior* limit to the maximum velocity of the wave that overthrew the object whose dimensions, &c., we have observed. And if any other regular solid, although dissimilar in form, can be found at the same locality, which has *not* been overthrown, we may obtain from it a *superior* limit of such velocity.

It is possible, however, that oscillation may occur to the limit of equilibrium, or even somewhat beyond it, without involving the fall of the body; for the relation may possibly be such between the time of oscillation of the body (Fig. 5) upon one of its edges or arrises f, and the time of a complete phase of the wave, that the equilibrium may be restored by the movement impressed in the second semiphase of the wave in the contrary direction to that first communicated, and before the body has had time to fall over, beyond the limit of such restoration; the adherence or friction of the arris f, with the base, producing the necessary hold, by which the wave so acts upon the body during its second semiphase, in the direction a' to b'.

If this be sufficient to bring back the centre of gravity through the horizontal distance between the verticals c and f during the time of the second semiphase of the wave, the body does not fall, but on the cessation of earth movement, topples back to its original position of perpendicularity, overpasses it, and after a succession of decreasing oscillations remains vertical as at first.

In so resuming a position of rest, it may be so circumstanced as to the nature of its base and arrises of oscillation, as to twist considerably from its first position, round one or more vertical axes.

This is a condition of things that very rarely occurs, except with small objects, like vases, statues, or pinnacles consisting of a single block. There are few masses actually found sufficiently hard, when of large size, to prevent the arris f splintering or crushing at the first movement to such an extent as to destroy all chance of restoration of position, even if the mass held together as one block; but in walls, towers, campaniles, or other compound masses, made up of blocks more or less firmly united, dislocation at several points takes place from the outset. Such masses being more or less flexible and elastic, *bend* first, break at the moment of maximum velocity of the wave, and then topple over piecemeal.

With the same velocity of wave, very different effects are produced, with regard to overthrow as the angle of emergence varies, and as the form of the body is different. Thus (Figs. 6 and 7), in the first, the wave emergent in the

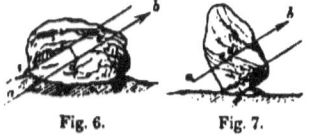

Fig. 6.   Fig. 7.

direction a to b, through the centre of gravity produces no disturbance of position in the "boulder stone," the extreme point s of the bed preventing rotation in the first semiphase of the wave, by a force measured by $d\,c$, and the point f in the second semiphase by a force measured by ef.

In fact, no velocity of earth wave occurring in nature, even with the emergent angle $e = 0$, i.e., horizontally, could overturn a block proportioned as in Fig. 6. If resting on a bed of earth or stone, it might slide and plough along upon it, and knowing certain coefficients, the length and dimensions of the channel or course cut by it would enable the wave velocity to be arrived at. In the case of the other block (Fig. 7), however, it would be overturned by a wave emergent in the direction a to b in either semiphase of the wave, the forces of overthrow in each being proportionate to $d\,c$ and $e\,f$. So in more regular solids, the column shaft (Fig. 8) may be overturned by a sufficient velocity of wave, in *either* semiphase, emergent at any angle between $h\,e$, passing through the centre of gravity and the horizontal wave-

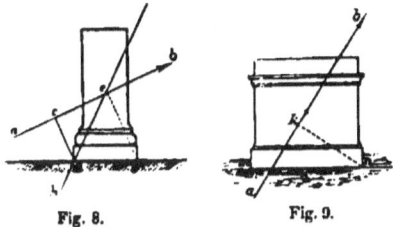

Fig. 8.   Fig. 9.

path passing through the same, the overthrowing force, with a given emergence and direction of wave-path, a to b, being proportionate to $c\,d$ in the first semiphase and to $e\,f$ in the second.

But the pedestal or "cippus" (Fig. 9) can only be overturned in the second semiphase of the wave, however great its velocity, if emergent in the direction a to b, nor then unless with a very great velocity, the effect of the wave in its first semiphase, however great its velocity, being merely to urge the whole solid against the ground in the direction b to a; and if it stand free upon a surface with friction (as an article of furniture, a cabinet or press, for example), to cause it to slide in the direction n to m horizontally.

The initial velocity of a body projected by earthquake shock, or that of some point of one overturned, is equal to the *maximum* velocity of the earth wave; for upon the principle of the equality of action and reaction, the greatest effect produced must be due to the greatest applied velocity.

And so also of fractures; they are to be considered as due to a force $M \times V$, M being the mass of the fragment broken off, and V the velocity of its centre of gravity or of oscillation, and equal to the maximum velocity of the wave, at the instant of its passing through which, fracture occurs.

It will thus be apparent that the principal phenomena presented by the effects of earthquake shock upon the objects usually occurring upon the surface of the inhabited parts of the earth, resolve themselves into problems of three classes, and are all amenable to mechanical treatment, viz.—

> 1st. Problems relating to the directions and amount of velocities producing fracture or fissures.
> 2nd. Problems relating to the single or multiplied oscillations of bodies considered as compound pendulums.
> 3rd. Problems referable to the theory of projectiles;

in which last, as the velocity is small, and the mass usually great in proportion to the range, which is also small, we are not disturbed by any consideration of resistance from the atmosphere.

These three classes of problems frequently are found combined in a single example—thus fracture and overthrow often occur together, or fracture and projection, and sometimes all three are united; a body (a gate pier for example) being broken off at its base, and overturned, but with a velocity more than sufficient for both, so that it is also projected, or thrown to a certain distance from its base. Although a less regular arrangement, it will tend to greater clearness, now to leave the further strict mechanical consideration of these questions, with the preliminary statements that have been made, and proceed first to describe pretty fully, the characteristics and details of structure of the buildings, &c., to which those principles will be applied in the present Report; and then to enter minutely upon the nature, of the fractures or fissures produced by earthquake shock in such buildings, &c., and describe the methods and conditions of observing them, and afterwards treating the observations; and finally to give in a connected form the equations referring to the treatment of all the classes of problems, as respects velocity, or direction obtained by calculation from velocity.

Almost all that follows with reference to the observation of the *directions of fissures* (or fractures), therefore is to be viewed as descriptive of the methods of arriving at *direction only* of wave-path, without reference to the velocity of the wave particles, or to any other mechanical conditions except those which determine the directions of such fissures as observed in buildings, *velocity* being determined, and *also emergence*, indirectly, by calculation applied to the observed conditions of the *forces* employed in producing fracture, overthrow, or projection.

Fissures in buildings, not overthrown, are, in fact, the sheet anchor, as respects direction of wave-path to the seismologist in the field.

The observations usually available, upon single or isolated objects, such as pillars, obelisks, vases, or statues overthrown, or others often of small size though comparatively rarer, are generally not less important, as determining direction, than those to be made upon objects of united or complex construction, or however large, such as buildings of various sorts. The latter are, however, the staple indices upon the careful observation of the injuries to which, we are mainly dependent for arriving at the directions in azimuth of shock.

CHAPTER III.

CONDITIONS OF EARTHQUAKE ACTION UPON ARCHITECTURAL STRUCTURES.

The effects produced by precisely the same shock, acting upon buildings differing in position, construction, material, &c., are so great, that it will be necessary to treat somewhat in detail of the conditions of earthquake action upon architectural structures; for without a thorough understanding of these, one is almost certain to be led astray by the strange, and often, at first sight, perplexing phenomena of destruction observable.

Throughout the kingdom of Naples, the edifices of cities, towns, and rural places present very uniform and striking characteristics, though varying much in dignity and size, &c.

In a few of the largest provincial cities, such as Potenza, Melfi, (No. 10, Coll. Roy. Soc.,) &c., the buildings, more especially those of government, the ecclesiastics, and the great landowners, present more or less of the majestic size, and architectural style, of the city of Naples itself. Loftiness, thickness of walls, apertures few but large, square-headed windows, and arched doors and gateways, with heavy tiled roofs, of low pitch, and with deeply overhanging eaves characterize the outside. The style of architecture, when style is attempted, is generally Roman, with cinque cento, or a still later and more debased style of ornamentation. The usually grandiose effect, however, very generally conceals, building workmanship of a very inferior quality.

The building materials of the kingdom generally, are lavas and tufa in the volcanic districts, limestone of various qualities, and brick (these are by far the most prevalent); and, in their respective localities, some sandstones, slaty rocks, and very rarely those from the ancient igneous rocks.

Limestone and brick are the staple materials of the regions to which this Report principally refers, except those of Naples and Melfi. The limestone is very seldom found, either in the jurassic or cretaceous formations, well bedded, or capable of being raised in long flat blocks. Lime is abundant, but the mortar often of very slender cohesion, from too great a proportion of lime and the want of a proper quality of sharp sand. Hence the general style of construction of wall, even in first-class buildings, consists of a coarse, short-bedded, ill-laid rubble masonry, with great thickness of mortar joints, very thick walls, without any attention to thorough bonding whatever. The opes of windows and doors often have cut limestone jambs, lintels, and dressings, which are but ill connected with the rest of the walls. In general, the external faces of the walls are concealed by plaster or rough cast. This is even the usual style of building for the better class of churches and monasteries. It has prevailed from a remote period, and a fair average illustration of its appearance is seen in the west end of the Villa Carusso near Auletta (No. 11, Coll. Roy. Soc.).

The floors in the better sort of town houses and palazzi, are formed of joists of fir timber, very commonly round as it grew, from 6 to 9 inches in diameter, placed at about 3 feet apart. The ends are inserted some inches into the walls, but are neither bedded on, nor connected by, any "tossils" or bond timbers, none of which are ever placed in the walls. Upon these joists a planking of fir, oak, or chestnut, from an inch to an inch and a half thick, is laid, rough as it comes from the saw, and pegged or spiked to the beams, and upon it a bed of concrete or beton, composed of lime, mortar, and broken tufa, brick, or stone is laid, to 6 to 8 inches in depth, and the surface of the latter is laid with red tiles—square or hexagonal—or sometimes plastered over with puzzolano mortar, and painted in oil.

The under surface of the floor is often bare, and the joists visible; in other cases a plastered ceiling is secured, by heavy lathing, up to the joists. See Photog. No. 12 at St. Pietro (Coll. Roy. Soc.)

A floor of this sort weighs from 60 to 100 lbs. to the superficial foot. Floors of palazzi, are also not unusually formed of arches and groins, built of hollow pottery embedded in mortar, the haunches filled in with beton, and plastered soffeits, with tiled surfaces to the floors, which, thus constructed, are of still greater weight.

The roofing also usually consists of round fir timber. The framing is of the simplest character except in some church and other roofs of great span, when the timber is squared. It consists commonly of principal rafters at 3 to 5 feet apart, connected by a rude collar brace, of round fir also, trenailed or bolted to the rafters. The feet of the rafters sometimes rest upon a wall plate of squared or of half-round timber, but often bed directly on top of the wall. These principals are crossed by stout sawed laths, and upon these are laid the common heavy ridge and furrow tiles, whose appearance is so familiarly characteristic of Italy, and so much more picturesque, than constructively good. These tiles are from ¾ to 1¼ inch thick, each course from 18 to 24 inches long, and they are frequently laid dry, and not secured down in any way but by their own great weight, except at the ridges, where the ridge tiles are cemented down in mortar. Roofing of this character weighs very little less than an equal surface of the flooring just described.

Framed roofing of large span and squared timber is not common in churches, &c., which are usually vaulted with brick or stone, dome'd or groined.

It will thus be remarked that in the construction of the more important buildings, the mass and inertia, of walls, floors, and roofs are enormous, while the bond and connection of each of these, and of all to the others, is loose and imperfect.

It is in the mediæval towns and villages of the interior provinces, however, that these conditions are still more evident. Nothing can be more striking than the general appearance of these ancient abodes. They are almost without exception perched upon the summits and steep flanks of precipitous "collines," usually rounded conoidal hills of limestone, sometimes abrupt and rocky elevations, whose slopes and shelves are occupied and their craggy heights crowned by the houses, built out to the very edge of the precipice, with no windows or doors looking outwards, or, if any, high up and inaccessible to any

who should climb the rock. Seen from beneath, in the valley bottom, through the keen bright air, and relieved against the sky, these old towns seem as though we could reach their interior in half an hour's scramble; yet often three hours' painful toil upon our mule will but suffice to bring us—by long traverses over rough and rolling stones, and by an approach road that is often the bed of a torrent in time of rain—to the ancient gateway, or to the narrow and obstructed street entrance by which alone we can penetrate the interior. Everything about these places is characteristic of their origin, its remoteness, and of the savage manners and times in which they were founded.

The irregular and narrow streets, not more than from 5 to 12 or 15 feet wide, are steep as staircases, until we reach the very summit of the town, where the little "piazza" and the principal church, or some gloomy-looking monastic pile, mostly form its centre and heart. We pass along between houses of all heights and sizes, beetle browed, and with low arched "portone," and small, unglazed, and often sashless windows, few, and high up. The unpaved and unformed surface, often the bare rock worn into steps, of these wretched streets, is the common receptacle of the filth of every house; pigs at all times, and often goats at night, make them their common resting-ground. There is neither sewerage nor water supply, and in winter wet, we wade through ordure ankle deep. Castelluccio (see Photog. 13) is a good illustration of the site and exterior of many of these towns. They all still retain the impress, of the semi-oriental character of the early settlers of Magna Græcia, of the savage violence and tyranny, of Saracen and Lombard conquerors, of middle-age superstitions and barbarism, and of a people condemned for ages, by misgovernment to an unprogressive state of ignorance and poverty, in the midst of the richest bounties of nature.

The towns owe their elevated position, primarily beyond doubt, to the necessity for defence and security in ancient times; but an universal belief exists that this elevation secures them against malaria, as it certainly relieves them in the summer from the unbearable reflected heat and pent-up air of the valley bottoms. These advantages, however, seem dearly purchased at the cost of difficult accessibility, even were proper road approaches made to them.

No roads whatever, suited to wheel traffic, exist throughout the kingdom, except the five great military ways, and these are perfectly unconnected by branches, with any but a few great towns: hence all produce has to be carried by mules, or by hand; and journeying off the military road can only be accomplished in the same way, or on foot.

It results from the perched positions, of almost all these towns that they are exposed to the severest effects of every earthquake shock. They are rocked as on the tops of masts. Padula is a good example of the larger and less ancient class of these towns (Photog. 14).

The style of building in these provincial towns, is much the same as has been already described of the cities, but poorer and humbler. The houses are seldom under two stories, rarely exceed three. The huts of the poorest classes (the land labourers, and shepherds) are but one story, huddled together in utter confusion; and the chief difference in point of masonry, in these country towns

PADULA

from that described is, that surface limestone—or that taken from the naturally exposed beds of rock—is commonly used to save labour in obtaining better, and hence the walls built almost invariably, of this coarse "nobbly" rubble, in half-rounded blocks, or rather lumps of stone, of nearly equal length, breadth, and thickness, and resembling nothing in form more than irregular loaves of bread, are almost devoid of masonry bond, and are shaken down into a heap, by a shock that would only fissure a well-built and properly bonded structure.

It results, too, from the extreme steepness of the scarps and terraces upon which these poor edifices are placed, that when some are shaken down they fall against and upon those that are beneath them, and increase thus the common ruin. This took place with dreadful effect at Saponara and elsewhere in the shock of 16th December, 1857.

The hill sites of these provincial towns are found most commonly on the summits and flanks of the lower spurs of hills that skirt the great mountain ranges, and are on the confines of the "piani," or great valley plains or slopes, that separate the chains; but sometimes they are absolutely upon lofty mountain tops (Conturso, Montesano), or at the edges of steep ravines (Bella); or on spurs high up on mountain flanks, as Petina, on the flank of La Scorza. Occasionally they stand (or stood) upon the flat tops, of insulated and enormously deep masses, of loose alluvium and clay, like Montemurro and Sarconi, with large rivers or torrents running at the bases of the clay cliffs, and eating them away.

This is almost universally the case in the great piano of Calabria Ulteriore Primo, and hence the expression of

Dolomieu, as to the destruction of the towns there in the great shock of 1783, that "the ground was shaken down like ashes, or sand laid upon a table."

Further remarks as to the situation of these towns, however, will best be made when observing upon some of the great physical features of the earthquake region, and of Naples generally.

With these remarks as to the general character of the buildings we have to deal with, I now proceed to the consideration in detail of the effects of earthquake upon them, and the phenomena presented by *fractures* and *fissures* in their walls, floors, &c. &c.

CHAPTER IV.

FIRST CLASS OF DETERMINANTS—FRACTURES IN RECTANGULAR BUILDINGS AS EVIDENCES OF WAVE-PATH.

IF an isolated wall (a parallelopiped) of masonry or brick, founded on level ground, be subjected to the transit of an earth wave, whose velocity is sufficient to affect the continuity of its parts, the resulting fractures will vary with the direction of the wave-path as respects the plane of the wall, and with the angle of emergence of the wave.

1st. If the wave-path be horizontal, or nearly so, and *in the plan of the wall*, the earth moving forward beneath the wall, tends to carry it forward by the grasp of its foundation and at its own velocity; but this is opposed by the wall's inertia. The material of the wall being, within narrow limits flexible and elastic, the tendency is to distort its figure, thus. The wave reaching the end *a* (Fig. 14 *bis*) first, with a transit from *a* towards *b*, the end *a* first begins to assume the form *e a*, rapidly taken by the whole wall, if sufficiently high in relation to its length. The wave traverses beneath the whole length, the materials, in virtue of their elasticity, oscillate in the same direction between *e* and *f* throughout the whole mass, and if the wall be fissured,

Fig. 14 *bis*.

it will be by a nearly vertical crack, widest open at top, and extending more or less down towards the base. If the wall were of absolutely uniform cohesion, there would be two such fissures near either end, or only one in the mid length, dependent upon the density, cohesion, and rate of force transmission of its materials, and the velocity of the wave movement: practically, such a wall is usually fissured in the weakest place.

2nd. If the wave transit be horizontal, or nearly so, and *oblique to the plane of the wall*, the latter either falls prostrate wholly, or a triangular fragment is thrown off from the end last reached by the wave, and in the direction contrary to its transit, or the wall is fissured, as in the first case only, dependent chiefly upon the greater or less obliquity of the line of transit to the plane of the wall.

Isolated walls, exposed to oblique or to directly transverse action, thus when tolerably thick, may sometimes be twisted considerably out of plumb without losing equilibrium or complete cohesion.

3rd. If the wave emerge *with a steep angle to the horizon*, the distortion is that of compression in the diagonal of the wall's plane, nearest parallel to the line of wave transit; and the fissures, if they occur, are also diagonal to the horizon, and approximate to directions perpendicular to the lines of pressure, *i. e.*, to the line of wave transit.

If the velocity of the wave be sufficient, in relation to the density and cohesion of the wall, a triangular mass may be projected from the end at which the wave passes out from it.

Reference will frequently occur to the directions in azimuth and emergence of the earth wave, relative to those of walls, buildings, or other objects affected by it. It will be convenient, therefore, to fix a nomenclature for these relations. A rectangular building, two of whose walls run north and south, and the other two east and west, may be called a *cardinal* building: buildings whose four walls run in any other azimuths will be described as *ordinal*.

Referring generally to the direction of wave transit in its horizontal component, or when nearly horizontal, as affecting cardinal buildings (which alone are generally suited for observation), it will be denominated *normal* when its azimuth is parallel to *either* pair of walls, viz., either north and south, or east and west.

When the line of wave transit, or its horizontal component, are in some intermediate azimuth, it will be said to be *abnormal*.

When a normal wave is an *emergent* one (the line of transit, or wave-path, inclined to the horizon) it will be called a *subnormal* wave; and in a similar case the *abnormal* wave will be designated as *subabnormal*.

These expressions will save much prolixity.

When the observer first enters upon one of those earthquake-shaken towns, he finds himself in the midst of utter confusion. The eye is bewildered by "a city become an heap." He wanders over masses of dislocated stone and mortar, with timbers half buried, prostrate, or standing stark up against the light, and is appalled by spectacles of desolation (such as those in Photogs. Nos. 15, 16, 17, 18, 19, and 20, Coll. Roy. Soc.).

At first sight, and even after cursory examination, all appears confusion. Houses seem to have been precipitated to the ground in every direction of azimuth. There seems no governing law, nor any indication of a prevailing direction of overturning force. It is only by first gaining some commanding point, whence a general view over the whole field of ruin can be had, and observing its places of greatest and least destruction, and then by patient examination, compass in hand, of many details of overthrow, house by house and street by street, analyzing each detail and comparing the results, as to the direction of force, that must have produced each particular fall, with those previously observed and compared, that we at length perceive, once for all, that this apparent confusion is but superficial.

We discover the cause, and in doing so obtain the key to all future correct and ready detection of the general directions of shock, by having learned to choose the proper class of buildings for our observations.

We find that wherever the ruin is complete and featureless—defying deduction—there the streets have been narrowed to five or twelve feet wide, have run winding hither and thither, ascending and descending, and that the walls of the houses, following their irregularities, have stood in every possible azimuth; that the exposed fronts and sides of the houses have faced every point of the compass; and often that the confusion produced by the shock thus reaching walls at the same moment at every conceivable angle, has been further increased by the falling houses having staggered against each other, and so some that might if alone have fallen in other ways, or might have escaped with only fissures, have been beaten to the earth by their neighbours. This sort of destruction, too, we will have remarked, belongs to the poorest habitations and worst built and densest parts of the town, where the wretched rubble masonry falls incoherent at the slightest jar.

We advance then to the churches, the barracks, or castello, the monasteries, the Casa Communale—to any of the better built and *isolated*, or nearly isolated buildings, and we soon discover that amongst these there are some that in every place present certain grand characteristics of partial or complete overthrow, and that these are everywhere generically much alike.

These we observe, aided by the prismatic compass, and with our previous dynamic knowledge, and soon discover that wherever such buildings, and not very dissimilar to each other, have been placed under like conditions, and so that their walls are in the same azimuths, like dislocations have affected them, and that where the directions in which the forces that we know *must* have produced the observed dislocations have not passed very diagonally through the walls, they have produced effects, regular and accordant with each other, and from which the directions may be inferred in which the forces themselves acted.

After a little experience we discover, that in every town (and frequently in other places) we may find rectangular buildings whose walls run very nearly north and south and east and west, and that these respond to our questions best; and finally, that buildings so posited, and having certain necessary characters of structure, when not too completely destroyed or overthrown, will enable us to discover the direction of wave transit, whatever may have been its line with reference to the walls.

It remains to describe, therefore, the effects produced upon such cardinal buildings by earthquake shock, to trace from the effects their causes—from the dislocations the forces that produced them and their directions, and to point out some of the more important modifications of effect due to differences of masonry, of form, of architecture, of wall apertures, and other such accidental conditions.

Commencing with the simplest case. If a *cardinal building* consisting merely of four unroofed walls be exposed to a *normal shock*, capable of fissuring the masonry, but not completely overthrowing it, the fissures will be found as nearly vertical cracks following the joints of the masonry, and within a few feet (more or less) of each quoin, as in Figs. 21-23, and Figs. 22-24, in plan.

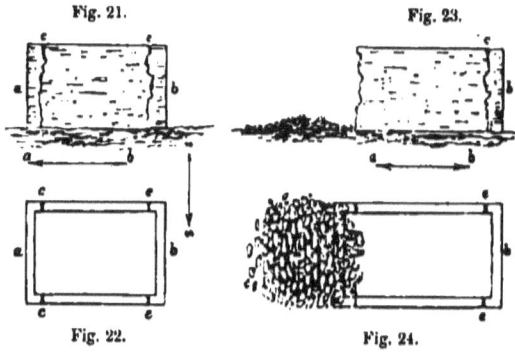

The fissures being widest at top, and becoming a scarce visible line at part of the way down the walls, or perhaps extending to their base, the earth wave, if in the direction *a* to *b*, reaches the end wall, *a*, first. Its inertia acts as an equal and opposite force at its centre of gravity, and tends to cause it to be left behind while the remainder of the building is pushed forward. The end wall *towards* the direction *from* which the shock has come moves in the opposite one, and if fracture occur the side walls fissure a short way off from the quoins, and the movement of the end wall is one of rotation round some horizontal line or lines situated along the length of its base.

Were every part of the walls of the building of perfectly equal coherence, and the rate of wave transit the same in its materials as that of the earth wave, the fracture would occur exactly at the internal angle of the wall at each quoin, breaking the side walls across in a plane coinciding with that of the internal face of the end wall. But the quoins are in practice built with larger, longer bedded, and better dressed masonry than the rest of the structure; and hence from this cause alone, without reference to others, the fissures are removed along the side walls nearer the middle, and into the less coherent masonry of the walls, and take place at c.

The earth wave pushes the side walls along with it, and these push the end wall b at either quoin before *them*. The end wall b therefore cannot fall by inertia in the same way as that, a, being propped up by the side walls. The earth wave, however, having passed its first semiphase, returns through the second half vibration in the opposite direction, and, we may assume, with equal velocity. The same set of forces now operate upon the end wall, b, the movement of the whole mass being in the direction b to a, and as described for the former end wall; so that b tends to turn over upon its base in the contrary direction to the movement of the wave itself in its second semiphase, and the side walls are fissured as before at a distance from the quoins greater or less along them, as at e (Fig. 21). As the force producing fracture and dislocation at any given velocity is always proportionate to M—the mass broken off or dislodged—so the extent of dislodgement after fracture (the materials being the same) is always proportionate to the velocity; and hence in any one building of like material and masonry the width of each fissure is proportionate to the velocity that has been effective in opening it; and we may compare component velocities in the directions of the planes of parallel or abutting walls by means of the widths of such fissures, the width in every case being measured with reference to an unit in length of the fissure from its origin, or where it becomes evanescent. This unit length may be arbitrary, but 10 feet in length of fissure is a very convenient unit, and the widths expressed in inches and decimals for that unit.

It is almost invariably found that in every building (with certain exceptions, to be noticed), although the masonry and form, &c., of the building may be quite or very nearly alike at both ends, the fissures c and e, do not occur at equal distances from the respective quoins (measured along the side walls), nor are they equally opened, large, and long, at both the opposite ends.

Whether this arises—as, from other considerations respecting the vibration of pendulous lamps set in motion by shock, and to be hereafter noticed (Part III.), seems probable—from a real difference in velocity in the two semiphases of the wave itself, and that the second semiphase is described with a somewhat slower velocity than the first, owing to defect of perfect elasticity in material substances composing the earth's surface—or whether it is due to the conjoint action of the elastic wave (the earth wave) itself, and of the wave of elastic compression of the materials of the walls themselves—or to whatever other cause, which future research must make clear, the fact may be accepted as certain and very general—that the end wall which is first acted upon by the wave (whenever it is something near normal), has the higher velocity shown upon it, and that the fissures at that end are, *cæteris paribus*, found to be wider than those at the opposite one.

The fissure formed at the end a, that first reached, is frequently rather wider than what is precisely due to this difference in velocity in the two semiphases of the wave ; for the end a is first fissured, the end b is next fissured by the second semiphase, which leaves the end wall b broken off, behind it, but carries back with it both side walls in the direction of its own return motion, and towards the end wall a. But more or less dust and broken fragments are often intercepted in the fissure a when first opened. These hinder the mass broken off at a from approaching the side walls, and closing the fissures (by the inertia of the broken-off mass), so that the side walls, in this return movement, push the end wall a before them, through the intervention of these obstacles, and so a second movement is impressed (small in extent) upon the broken-off end a, in the contrary direction to the wave transit, and in the same direction with its first movement, which ends by increasing the final width of the fissures at the end a.

The chief disturbing causes that interfere with and mask the regularity of this phenomenon are—The wave being subnormal and emergent at a considerable angle ; in which case the length of the dynamic couple (as has been already generally explained) that measures the overthrowing and dislocating power at a given velocity, is greatest in the second semiphase of the wave, and to such an extent, as to obliterate the effect of the difference in velocity of the two semiphases, and either leave the fissures equal at both ends, or even make those at the end b the wider.

Inequalities in the materials or masonry at the opposite ends in the line of the wave-path—want of complete or nearly approximate symmetry in the size and form of those ends—perforations of doors or windows, or suchlike sudden changes of continuity of wall—and great length of bond in the wall stones at all or at a few points—are the other conditions which chiefly perplex and interfere with the phenomenon.

Practically, however, this fact is a guide of much importance in seismic observation, inasmuch as it enables us very frequently to decide, with more or less certainty, as to the direction of wave transit, from conditions that otherwise would afford no information beyond that of the path of the wave, leaving it quite uncertain whether the seismic vertical were to be sought for towards a or towards b.

Where the phenomena are clear, we may, on the contrary, always conclude that it lies along the wave-path, *towards the end that presents the widest fissures.* Very few large and massive cardinal buildings will be found that will not give, as respects a normal or slightly subnormal wave, a decisive response, from some or other of its parts, by this means.

The actual phenomena in a well-developed case are illustrated in the Photog. No 25, which shows the front end of the church of Pertosa, looking at its N.W. end. The direction of the nearly horizontal wave that pro-

CATHEDRAL OF MARSICO NUOVO.
North Side.

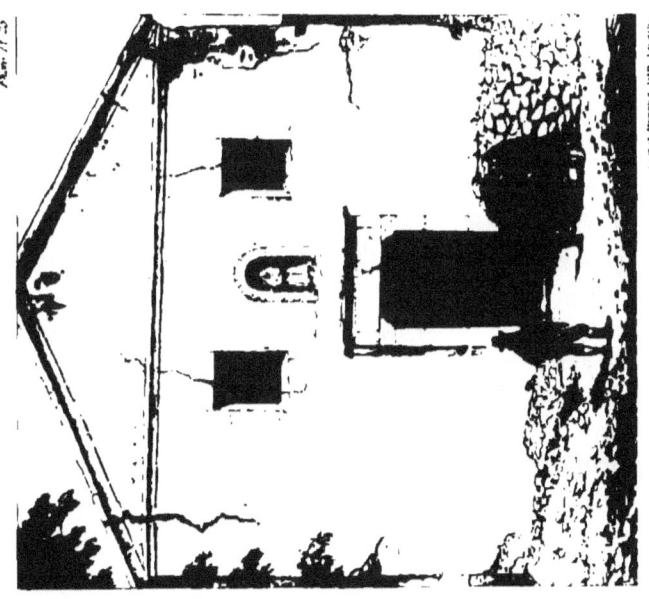

CHURCH at PERTOSA, *looking North West.*

duced these fissures having been almost parallel to the west end wall, and from the N. E. to the S.W., or from left to right of the picture. The wave of shock was of very steep emergence at Pertosa; but the whole "colline" upon which the town is perched, oscillated laterally with the wave, and horizontally, or nearly so. The horizontal fractures, and those over the door, are due to the emergent wave only.

The force producing fracture and dislocation, impressed by the shock, may be viewed as separated into two—one just sufficient to fracture the materials, the other to dislodge them more or less. Both depend upon the velocity at maximum of the wave; but the power to produce fracture depends much more upon velocity than upon the amplitude of the wave, while the energy to produce dislocation after fracture depends also upon the latter, which determines the time during which the motion of the passing wave acts upon the mass.

The flexibility and elasticity of masonry or brickwork, even of the highest quality, in masses of ordinary size is small, the limits of distortion without rupture narrow: the compressive or extending forces being due to inertia, are proportional to M V, and for the same material proportionate to V only; and as the amount of extension or compression, for the unit of length due to any force *suddenly* applied to an elastic solid, is double that produced by the same force, if statically or slowly applied, the effect of a high velocity is to produce fracture with great facility in bodies of narrow elastic limits. A wave shock of extremely small amplitude, therefore—one so small as not to appeal at all alarmingly to our senses—may yet be competent to produce considerable fractures in buildings; but in this case the fissures will be found to be close and thread-like.

Were the mass of a wall (viewed as a single parallelopiped) so circumstanced that, its integrant portions, retaining their relative positions merely, were free to oscillate, and then to remain at the points which they occupied at the moment the wave left them, having no resilience, in such case the chord of the arc of movement at the centre of oscillation would be very nearly equal to the amplitude of the wave that produced the oscillation; and this would be equally true of the width of a fissure produced in such a wall. Now, we occasionally find walls that are extremely massive in proportion to their altitude, and built of small stones or brick laid in bad masonry, and with almost bondless mortar; such walls have little or no resilience, and, when thrown more or less out of plumb, or fissured, with suitable conditions, afford a rude approximate measure of the horizontal amplitude of the earth wave, by the range of movement impressed at the level of the centre of oscillation. Some examples of this, as observed, will be found in Part II.

The *adhesion* of mortars and cements to stone or brick, in a direction perpendicular to the faces of the joints of the work, is always much less, with the ordinary materials employed, than the *cohesion* of the latter for equal sections; the exceptions being only buildings of very soft tufa, or some such stuff. Hence, although fracture and open fissures may occur occasionally, running right through some stones and breaking them across in a building which may be acted on transversely, as when very long upon their beds, and crossing a line of fissure near the axis of revolution, thus (Fig. 26); it nevertheless almost invariably happens that the line of fracture, whether in stonework or brickwork, follows down or along a line of joints, producing a jagged or serrated fissure, the jaws or serrations depending upon the length of bed of each block or brick, and the depth of the courses.

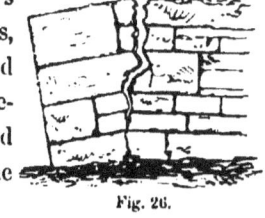

Fig. 26.

It has been found that the adhesion of Portland cement to Portland stone is only 146 lbs. per square inch, while the cohesion of the cement itself is 400 lbs. per square inch, or the former little more than one-third; and that the adhesion of Parker's cement to granite is as low as 22 lbs. per square inch, the cohesion of the cement being 300 lbs. per square inch, or less than one-thirteenth. The adhesion of common lime mortar varies enormously, with the nature of its materials, the sort of stone or brick which it is used to cement, the thickness of the joints, the care taken to fill them effectually and solidly, the degree of wetness or dryness of the mortar itself and of the stone or brick to which it has been applied, and the rate at which the mortar has been dried during its setting, and the amount of moisture and of air to which it has been subsequently exposed. All these conditions, or some of them, have been found sufficient to make a difference of absolute cohesion of more than 2 : 1 between old Roman mortar consolidated and hardened for ages, and good modern mortar allowed sufficient time to be viewed as fully set or indurated. When very dry, mortar is much more brittle and easily fractured than when wet even after complete induration.

Such conditions, and others similar, but tedious to detail, must be known to, and constantly looked out for, by the seismic observer in the field, or otherwise he will continually be liable to compare, as to effects, unlike and incomparable buildings or circumstances.

The coefficients of cohesion which apply to our equations of fracture will be given hereafter.

Returning now to Figs. 23 and 24. If the path of the wave be normal as before, but its velocity and amplitude greater than are sufficient only to cause fissures, then one or both end walls may be overthrown. If the direction of the transit be from a to b, the end wall a will be prostrated outwards, or in the contrary sense. The end wall b, propped, as before explained, by the side walls, may *possibly* be projected outwards and fall also; but in most instances there will be only fissures produced at its end, as in Figs. 21, 22.

This may be made clearer by referring to Fig. 27. Let A E be the path of the wave, its direction of transit being

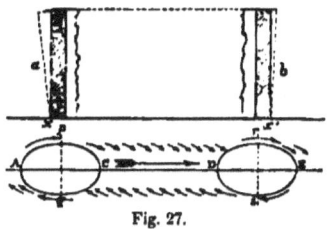

Fig. 27.

from A towards E, and the form of the wave vibration cut by a vertical plane, be A p C q at the end a, and the same when it has progressed to that b; bearing in mind, however, that the amplitude of the wave, as it actually occurs in earthquake, is very great in proportion to its altitude, in most cases, indeed in every case of a normal or nearly normal wave, and that during its transit the whole building is simultaneously in motion.

The end wall begins to be affected by its own inertia at the moment that the forward phase of the wave A to C reaches it. The velocity of the vibrating mass increases to the maximum at the point p; when whatever fissure may take place occurs, and the centre of gravity of the mass begins to move in the direction from C to A, the whole turning round the point x at the base. This movement is continued, though with diminished energy, by the wave motion during the second half of its *first* semivibration, *i.e.*, from p to C, when it passes through zero, and now, during the whole of the second semi-vibration from C to A, passing through the second maximum at q, the motion of the earth is in a contrary sense to that of the wall, and of the wave transit.

It has set the detached mass in motion with a momentum $= MV$, V being the velocity of first semiphase of the wave itself at its maximum. It tends to destroy this during the second half vibration, by a momentum $= M(V-v)$, v being the difference of velocity in the semiphases. During the time of the second half, of the first semi-vibration from p (when fracture occurs) to A, the wall continues to fall or turn over outwards, and for a little beyond this; but this is now checked by the return or second semi-vibration, and unless the angular motion of the mass in the time from p to A shall have carried its centre of gravity beyond the vertical passing through x, the wall shall not fall. If such be the case—*i.e.*, if the wall do not fall—a contrary motion, more or less tending to restore its position, is impressed upon it in the return from C through q to A, and it comes to rest, with the fissure somewhat closer than it was, at an intermediate moment just after its formation, unless fragments have fallen between and prevented this, and always assuming that its parts hold coherent. Proceeding now to follow the train of action upon the end wall b, the wave affecting almost simultaneously the whole building, the two side walls and the end wall b are *forced forward together*, the movement as before, commencing at the instant the initial movement of the wave D reaches them. They both (side and end walls) pass through the point r of maximum velocity nearly together, and so to E, when the motion of the wave itself is zero, and the motion of its second semi-vibration commences, which is retrograde as before.

Fracture cannot occur at the end b during the first semi-vibration D r E, because the side and end walls are alike urged forward together and at equal velocities: there is therefore nothing to produce separation. If fracture, therefore, take place at the end b, it must occur at the point of maximum velocity s, in the second semi-vibration, from which to D, the motion of the wave continues to promote separation; but the momentum impressed at s is $= M (V - v)$, whereas at the former end at p it was M V. The force necessary to produce fracture of the materials being the same at both ends (which is, however, only strictly true for absolutely equal velocities), the amount of movement impressed upon the mass at the end a, will be greater than that at the end b, by the momentum due to M v, (neglecting any small restoration of position of a, at the return semiphase of the wave), and so if fissures only be produced, those at *a* will be wider than those at *b*, as was before stated; or if the impressed movement be capable of more than this, the end wall *a* will be prostrated; and that *b* may stand, but fissured from the side walls, or with still greater violence, it, too, may be thrown forwards in the same direction as that of the wave transit, but to a less horizontal distance.

Were the building in Figs. 21-23 square in plane, instead of rectangular, it will readily be conceived that precisely the same phenomena must succeed to a normal wave, whose path should be orthogonal to one in the direction *a ... b* or in *s ... t*.

If, however, the building be rectangular, and with the sides *c e* and *h k* of considerable absolute length, and largely exceeding that of the ends, and the path of a normal wave be through them in the direction *a ... b* (Fig. 28), it then

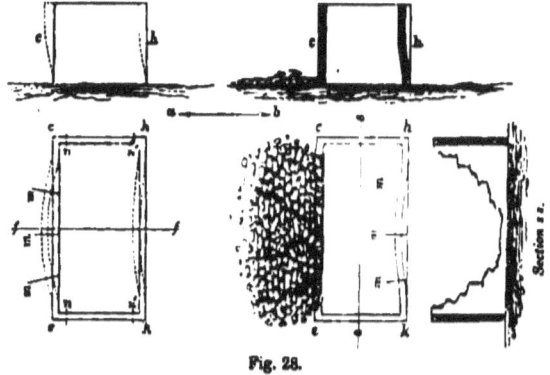

Fig. 28.

rarely happens that fissures occur, or occur alone near the quoins.

The wave, as before, passes from *a* towards *b*, and the side wall *c e*, as before, moves by inertia in the contrary sense; and were it sufficiently rigid along its length, it might tear off from the ends, and fissures occur as before at $n\,n$, $n'\,n'$, the other side wall $h\,k$ following the movement already described for b (Figs. 21-23).

The whole length of $c\,e$ has an equal velocity impressed; it is unsupported for its whole length, except at the two extremities, where it is connected by the quoins with the end walls, and held fast by them. It bends into a curve, therefore, along its length, bowing outwards most at the top and centre of length, and receiving several fractures $m\,m\,m$, approaching to vertical in direction, owing to the length of the curve being greater than that of the originally straight wall. The whole bond of the materials is more or less disturbed; but the force of the shock may only be such, as to thus curve and fracture, but not overthrow the wall. The wall $h\,k$ in like manner is urged forward at the ends h and k by the connection with the quoins of the end walls; but failing also in rigidity, the central part is left behind, and bent also by inertia; differing from the first case analyzed in this also, that the direction of movement impressed is not in the same direction with the wave transit, but reverse to it; so that here both side walls $c\,e$ and $h\,k$ move alike in direction, but to different extents. The greater velocity V is common to both, for both have their velocities impressed by the first semi-vibration of the wave; and were the nature of the connection at the quoins the same, whether the walls were forced outwards or inwards, both would, *cæteris paribus*, be bowed alike. From the nature of the quoin bond of masonry, however, the end walls at the quoins offer much less resistance to the wall $c\,e$, being forced outwards, than they do to the wall hk, being forced inwards, the quoin stones bonded into both walls finding a better fulcrum in the transverse resistance of the end walls in the latter case. The wall ce in some degree resembles a beam merely *supported* at the two ends, while that hk is partially in the condition of one *encastré* at both ends, both being subjected to transverse strains; and accordingly the latter usually presents the characteristic curve of double curvature when looked at in plan upon top, as in Fig. 29.

The difference in result is practically not great, but sufficient generally, to cause the side wall first moved by

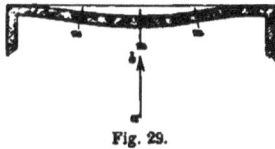

Fig. 29.

the wave, to have a greater curvature than the opposite one, both being more or less fractured. When the shock is of sufficient force, however, the wall ce is quite overthrown, falling outwards, and tearing away from the end portions, in an irregular, sloping, and hollow curve, following along the joints, so that the detached mass is in the form in the section in line ss looking towards ce. This seldom happens, when the wall hk is devoid of intermediate support, without its being more or less prostrated likewise. This form of fracture is seen in two of the walls in the Photog. (No. 30) of overthrown houses at Polla, though unfortunately somewhat masked by the effects of the falling walls having been in this instance precipitated upon and against others.

E 2

CHAPTER V.

FRACTURES CONTINUED—RECTANGULAR BUILDINGS—SUBNORMAL SHOCK.

We proceed next to consider the effects of a *Subnormal wave*, or one whose transit is emergent at an angle to the horizon, and in a vertical plane, parallel to two of the walls of a rectangular building.

If the angle of emergence—viz. that contained between the line of transit of the wave and the horizon be small, not exceeding 10° or thereabouts—the effects, when producing fissures or overthrow, can seldom be distinguished alone from those of a normal wave. Fissures are produced near the quoins, at the tops of the walls, and if overthrow takes place the end walls and detached portions of the sides, are thrown outwards as already described. In accordance with the general law, the fractures tend to place themselves at right angles to the direction of wave transit. Thus referring to Fig. 31, where the angle of emergence (of the wave whose direction of transit is a to b) is $h\,i\,a$, the end wall c is thrown back by inertia, and that at e projected forward, with a difference of velocity $= v$, as before. As the joints of the masonry or brickwork, which the fractures follow, upon the whole run vertically and horizontally, and as the fracturing force is transmitted diagonally in the direction $a\,d$ through the side walls, fracture occurs in jagged lines in directions perpendicular to $a\,d$, the wave-path. The fissures at the end c, therefore, commencing at top, very near the internal angles of the quoins, run down in the direction $p\,k$, making an angle

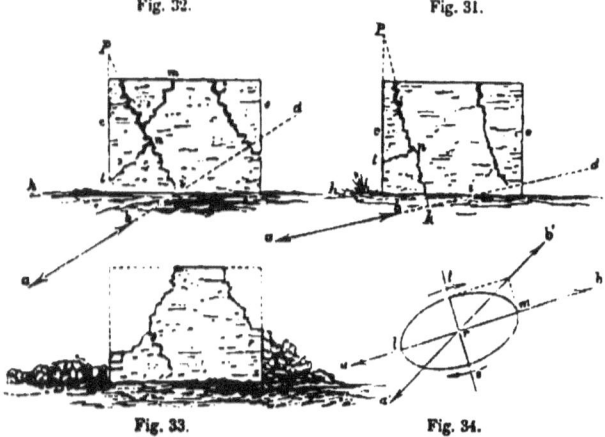

Fig. 32.   Fig. 31.

Fig. 33.   Fig. 34.

with the plumb line of the wall c (before disturbance) $c\,p\,k = h\,i\,a =$ the angle of emergence.

The portions of the side walls detached with the ends grow wider as they approach the base. They therefore make the end wall c too rigid, to turn round upon or near its base, as in the former case (Fig. 21), to admit of the fissure $p\,k$ opening; hence a cross fracture occurs somewhere below one-third the height of the wall c, as at n, which frees the mass and admits of its movement. This cross fracture may take any direction downwards from n towards t, dependent upon the nature of the masonry, &c., and the mass c chiefly turns over, to the extent of opening of the fissure, round the joint at t, or may partly slip upon the fractured joints in the direction da and partly turn. When the building is large, and the angle of emergence large also, as in Fig 32, this cross fracture is usually a curved line, more or less hollow downwards, passing out at the quoins at a bed joint as at t, and in such a direction that a chord to the hollow curve from n to t approaches to right angles with $p\,i$.

The end e (Fig. 31) is projected forward towards d by inertia in the second semi-vibration of the wave. Its fissure is more or less exactly parallel with $p\,k$ for the greater portion of its length, but it seldom runs to the base of the wall s, turning out towards d by a horizontal joint, somewhat above that level, and at a higher point, as the angle of emergence is greater, as may be observed in Fig. 32.

The reason of this is pretty obvious. The inclined direction of the fissure causes it to reach the internal angle of the quoin before it comes down to the base of the end wall e, which therefore breaks at that level along an horizontal line, and all below that, not being detached and loaded with the mass above, is unmoved. Cross fractures may or may not, follow from this fissure towards c, dependent on the breadth of side wall, occurring between the fissures at the ends c and e, and upon the angle of emergence, class of masonry, and other conditions.

In Fig. 32 the effects are shown of the wave when emergent at a still greater angle. The train of phenomena is quite similar to that just described, but with this addition, that where the angle of fracture, $t\,p\,i = h\,i\,a =$ that of emergence, is great, and hence the angle $i\,p\,m$ great also, the overhang of the upper part of the side wall, coupled with the momentum in the direction $i\,p$, due to the small motion of the wave in *altitude*, produce a fracture somewhere at *m*, or more than one, the direction of which downwards is much modified by the joints, &c. &c., of the masonry, and is generally nearer to the vertical than exactly at right angles to *p i*.

When the emergent wave produces a sufficient shock for complete overthrow of the end walls *c* and *e*, they fall as in Fig. 33, leaving the fractures of the side walls modified, by the grind of the descending masses. It is rarely, however, that a shock of an emergent wave, sufficient to throw back one end, and forward the other, occurs without the side walls being also thrown, in or out, or both, either by transversal wave motion or by secondary actions of the falling end walls upon them.

It was stated above that the direction of diagonal pressure through the walls was in that of the wave transit: this is perhaps not strictly true, for referring to Fig. 34, if *a* to *b* be the path and direction of wave transit, and the velocities of the wave itself be equal in altitude to *s t* and in amplitude to *l m*, then the resultant pressure due to its movement in the forward half of the first semi-vibration is in $a' b'$, which, combined with the motion of transit *a b*, will give a resultant pressure somewhere between b' and *b*. As, however, the amplitude *l m*, of the earth wave, appears to be always very great with reference to *s t*, there is no practical error (and much convenience for calculation), in considering the line of pressure as coincident with that of wave transit.

If the subnormal wave be orthogonal to that just described, and still affecting a rectangular building, so that its transit passes through the longer sides, these are bowed (if of sufficient length, &c.) outwards, at the side first reached by the wave, and inwards at the opposite one, differing in nothing from the effects of a normal wave upon the same building except that these are all less marked, for the same velocity and amplitude, the *effective velocity* producing movement in the masses detached, being to that of a normal wave of equal velocity, as $r : \dfrac{v}{\sec e}$, e being the subnormal angle, or angle of emergence.

Two other conditions require notice, however, as also affecting the dislocations produced by subnormal waves.

If such a wave, with a given value for the angle e, be resolved into a vertical and horizontal component, it is the latter that is chiefly effective in producing dislocation when e is small, the former when it is very great: and the effects of both are modified more or less by the form of the individual blocks of stone of which the wall consists. If these are very long in their beds, they offer a most powerful resistance to fracture or dislocation by a steeply emergent wave; and when thus long bedded, close jointed and squared ashlars, prevent any indication of value being had.

In the choice of buildings, therefore, for arriving at the value of e from subnormal fissures, those must be selected that are of large size, with walls of brick, or of rubble masonry of inferior quality, or at least of small, short-bedded stones in proportion to the size of the walls; and fortunately (for seismic researches) there is no want of such in the south of Italy.

Where, also, the value of e is great, two other circumstances come into play to modify the widths of the fissures, and even affect their direction, more or less.

Referring again to Fig. 32, the effect of the vertical component of the subnormal wave, in its first semiphase, tends to drive the fractured mass at the end *c* down upon its foundation, and to throw that at *e* into the air.

Gravity, therefore at the former end, acts *with* the wave, in the first semiphase, but *against* it, at the latter end of the building, and *vice versâ*. But the rhomboidal mass of wall between the fissures at the ends *c* and *e* is also acted on by gravity *with* the wave, and the result is frequently to force down great wedges, such as *p n m*, which close the fissure *p* in such a way as to prevent a certain indication, for these wedges being detached all round, remain where they descended to last. This is, however, a result of less importance, because there never can be a mistake, as to the direction along the wave-path of steeply emergent subnormal shocks, in which the seismic vertical is to be found; *it must lie to the side at which the wave-path dips* below the horizon.

The overthrowing power, and, to a certain extent, the fracturing power of a subnormal wave differs in the first and second semiphase (without reference to the difference due to difference of velocity), being proportionate to the perpendiculars to the wave-path, let fall from the centres of oscillation to the fulcra round which the fractured masses turn.

This is greatest at the side *e*, towards which the wave travels, and the tendency of this is to equalize the widths of the fissures. Other consequences will be apparent to the mechanical reader on considering the conditions.

From observation of the effects of a subnormal wave, therefore, we may be enabled to arrive at conclusions as to—

1st. The path of the wave—Subnormal.

2nd. The direction of transit—the fissures being occasionally most open at the end first reached by the wave, and *the transit always from the end that dips* below the horizon.

3rd. The angle of *emergence of the wave* with the horizon being *equal to the angle made by the main fissures (or those transverse to the wave-path) with the vertical.*

4th. The velocity of the wave motion may, under favourable circumstances, be inferred from that impressed upon detached and fallen masses.

Abundant examples will occur, in the second part of this Report, of these subnormal waves, and of their effects.

CHAPTER VI.

FRACTURES CONTINUED—RECTANGULAR BUILDINGS—ABNORMAL SHOCK.

WE have next to refer to the *abnormal wave*, or that passing horizontally, or nearly so, and diagonally through a rectangular building.

In this case the fissures occur at or near the internal angles of the quoins, and are vertical or nearly so, and in so far are the same as those produced by a normal wave. In a rectangular or square building they are also (the main fissures), the same in number (four) generally, but differently disposed. If the abnormal angle (that which measures the horizontal obliquity of the path of the wave, with two of the parallel walls,) is very small, as in Fig. 37, it occasionally

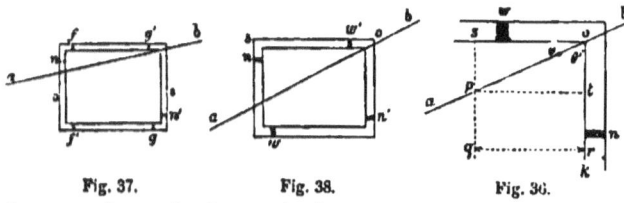

Fig. 37.   Fig. 38.   Fig. 36.

happens that only four main fissures are seen, the alternate ones g' and f' being rather wider than the two others, when the direction of the wave is from a to b, and the sum of $f' + f$, being greater than that of $g' + g$, on principles already explained. And sometimes these are accompanied by a much smaller main fissure at n, and either none corresponding in the end e, or one at n' still smaller. In such a case if the abnormal angle be less than 10° it scarcely admits of decisive observation.

When the abnormal angle, however, is greater, as in Fig. 38, four main fissures are formed, which, except in the case where that angle is = 45°, are alternately wide and narrow. Let the direction of the wave transit be a to b, making an angle (in an horizontal plane) less with the wall $o\ s$ than with the wall $o\ n'$. Then at the end n at which the wave first arrives, the fissure n will be narrow, and w will be wide, and at the opposite end, n' will be narrow and w' wide, and the sum of $w + n$ will be greater than that of $w' + n'$, all being measured horizontally across the jaws of the fissures at the same level, suppose at the top of the walls.

The cause of this is pretty obvious. Referring to Fig. 35, if the direction of the wave be a to b, the force of dislo-

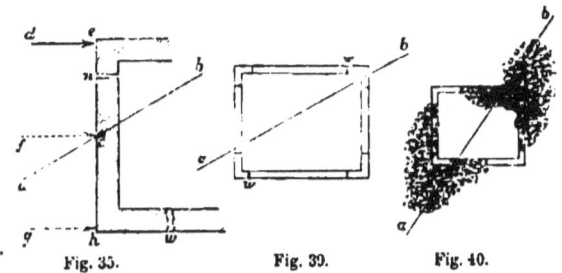

Fig. 35.　　　Fig. 39.　　　Fig. 40.

cation of the end wall $e\ h$, and the side $h\ w$, acts in the direction $b\ c$, through the centre of gravity of the end wall, and oblique to its plane; it is therefore resolvable into two, one perpendicular to the plane of the wall, $c\ f$, the other parallel to its plane, *c h*, and both through the centre of gravity or of inertia.

The component *c h* produces the fissure *n*. The resistance of the base of the wall, *e h*, in the directions *d e* and *g h*, uniformly along its length, may be resolved into a parallel force at *f c* in the centre of the length and at the level of the base, which is below the the level of the component force, *c f*, through the centre of gravity, there is therefore a dynamic couple, tending to turn the wall to the left round *e h*, and this produces the fissure *w*.

The fissure *n* may take place anywhere along the wall *e h*, but usually occurs towards *e*, at a distance from *h*, that is to the distance of *w* from *h*, approximately as the component *f c*, is to that *c h*, so that a line drawn across from *n* to *w*, either at the inside or outside faces of the walls, will cross the path of the wave at right angles more or less exactly. The position of *n*, however, is subject to many disturbing circumstances, and it never occurs (as the proportion of the figures might seem to infer) so far from *h* towards *e*, that the masonry of the wall *e h*, cannot yield sufficiently to the shearing strain in the direction of its length, to admit of the opening of the fissure.

Now from the observation of the widths of these respective fissures, we are in a position to find the angular direction in which the path of the wave has traversed the building. For referring to Fig. 36, let *w* be the wider and *n* the narrower fissure, whose widths of ope are proportionate to the components *a* and *b*; the path of the wave, being *a* to *b*, θ and θ' the angles (together = 90°) made by these components with it, and which we require to find, then—

$$\sin \theta : \sin \theta' :: a : b,$$

a and b being equal or proportionate to the widths of n and w, therefore—

$$a : b :: 1 : \tan \theta',$$

or

$$b = a, \tan \theta'$$

hence

$$\tan \theta' = \frac{b}{a};$$

from which either the angle θ' or its sine can be had from the tables.

As the path of the wave divides the angle made by the two walls (whether the building be square or not) in the ratio of

$$\sin \theta : \cos \theta, \text{ or of } \sin \theta : \sin \theta',$$

its direction may be easily got geometrically, by an observer not accustomed to trigonometry.

It sometimes happens that eight main fissures may be observed, alternately wide and narrow, in a rectangular building exposed to an abnormal wave. This only happens when the building approaches a square in plan, the abnormal angle being not far from 45°, and the walls very uniform in structure and mass, and built of small material, such as brick.

There are then two fissures, one wider than the other, near each quoin, as in Fig. 39; four are primary and due to the direct action of the wave, the other four seem to arise from the shearing strains in the plane of the respective walls. The case is unusual, but when met with is best rejected for seismometry, as likely to lead to error.

The walls are sometimes, though rarely, found as in Fig. 42, the middle portions of c and d first reached by the wave, being overthrown inwards, and those a and b leaning outwards. This mostly happens from quoins of stability disproportionate to the rest of the building, and unfits it for seismometry.

When the force of shock of an abnormal wave is sufficient to cause prostration of the walls, they almost always fall outwards, and the debris is found as in Fig. 40, *a* to *b* being the direction of the wave.

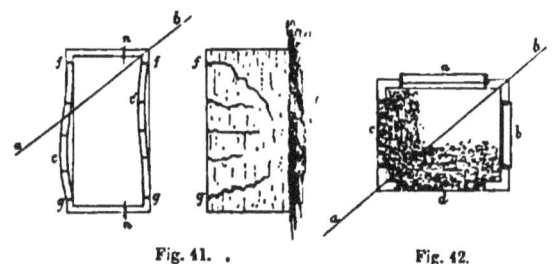

Fig. 41.   Fig. 42.

When the building is rectangular, and the abnormal wave in the direction *a* to *b*, Fig. 41, arrives first at one of the long side walls, making an angle of 45° or less, with the end walls, the latter are generally fissured vertically at *n n*, but the long side walls are also bowed, or possibly prostrated; the greatest amount of curvature being at *c* and *c'*, and the fissures taking the hollow curved forms, seen in the elevation of the wall *g f*, the central fissures, being secondary or sub-fissures, dependent upon the bowing.

This form of building is difficult to obtain the abnormal angle from, with correctness, and those more nearly square should be sought for.

The remarks that have been made apply to cardinal and ordinal buildings alike, the former, when presented, being by far the best, however, for observation.

We are therefore enabled from what precedes, in the case of a *cardinal building and abnormal wave*, to infer—

1st. The *path* of the wave.

2nd. The *direction* of transit *motion*.

Measures of velocity, can scarcely ever be obtained from an abnormal wave, as the overthrown masses are quoin pieces, which fall attached at right angles, and usually defy attempts to ascertain the moments of inertia and dimensions of base.

If the fissures be *clean and well defined*, and the walls not too much perforated by openings, or otherwise rendered irregular, good results can be had; it is then, not important that the walls making angles with each other, from which the widths of the directing fissures are taken, should be of equal thickness, because the force acting on each is proportional to its mass, and the section of fracture for equal height is so likewise; but they must be of similar material and masonry. All such conditions, however, will be better understood after we have treated of the perturbation of phenomena produced by architectural and other features, &c.

CHAPTER VII.

DIRECTION OF FRACTURES IN RECTANGULAR BUILDINGS BY SUBABNORMAL SHOCK.

WE now proceed to the fourth of our classes of waves, namely, the *subabnormal*, or the wave whose direction of transit, is *diagonally in both an horizontal and a vertical plane, passing through the building; a wave which is at once abnormal and emergent.*

On referring to Figs. 43 and 44 the general character of the dislocation produced by such will be evident.

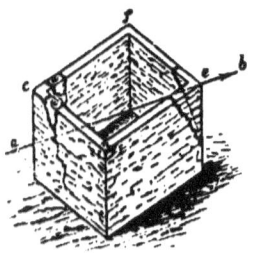

Fig. 43.

The wave emergent in the direction *a* to *b* dislodges the portions of the quoin *c*, which it first reaches by inertia, during the time of the first semi-vibration, and those of the diagonally opposite quoin, are thrown forward and often projected out of place also, by inertia in the time of the second half vibration. As explained for the abnormal

corresponding to the subabnormal sought; and the path of the latter will be found in a vertical plane passing through

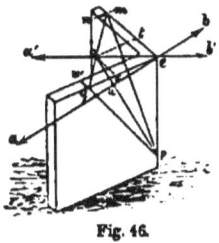

Fig. 46.

a' b'. Join $o\,p$, which is in the same vertical plane, and also in the plane of the fissures or fractures, $m\,p$, $q\,p$, and through the point of the quoin e intersecting $a'\,b'$, draw $a\,b$ perpendicular to $o\,p$; $a\,b$ is then the path of the subabnormal wave, emergent in the direction a to b. This is tantamount to finding the resultant of all the parallel forces that resisted fracture, and of course assumes, that the masonry fractures equally readily everywhere.

This is practically sufficiently near the fact, except, perhaps, when the horizontal obliquity of the wave-path, or abnormal angle is very great; in that case one wall is broken by direct pull nearly, and the other nearly transversely, which may give rise to an unbalanced couple at u, in a line parallel to $q\,m$; and in that case the wedge-shaped mass, in place of simply sliding down or turning over, in a plane passing vertically through $a'\,b'$, and falling to pieces, at the base of the quoin, will have a small amount of rotation, either to the right or left of that plane; and the centre of gravity of the mass of debris, will be found correspondingly posited, to the right or left of the base of the quoin. No case has been observed by the author (amongst very many of the class), in which from this cause a perturbation was produced, that could render the determination of the wave-path uncertain in the horizontal element, by more than 2° or 3°.

The arc corresponding to the sine of $t\ e\ o$, or of $s\ e\ o$, viz. θ and θ' of a former case, gives the abnormal angle or bearing, of the horizontal element of the wave-path, if the building be cardinal: or this, + or − the azimuth of the walls, gives it if ordinal, and the angle $o\ e\ u = \theta''$ is equal the angle of emergence with the horizon, to which it is alternate.

The whole of the preceding may be readily done trigonometrically, and by an observer accustomed to such operations that method will be found more advantageous, as greatly economizing time on the ground, and enabling the results to be worked out at leisure.

Let the dark lines $m\ e$, $q\ e$ be the level top of the adjacent walls (Fig. 48 *bis*), $e\ f$ the quoin, e the solid angle at top, n, w, p, the points of fracture in those lines.

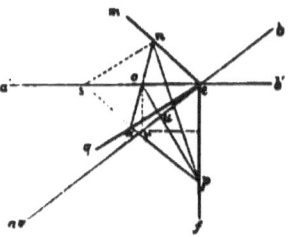

Fig. 48 *bis*.

As $e\ f$ is plumb, $m\ e\ f$, and $q\ e\ f$, are each 90°. Let $q\ e\ m$, *any* angle be given, and also the distances, $n\ e$, $w\ e$, $p\ e$, the two former proportionate to f and f', the component forces that produced the fractures at n and w; f'' the vertical component corresponding to these; u the intersection of the polar (or direction of emergence of the wave), with the plane passing through n, w and p.

The angles made by $o\,p$ and $a\,b = 90°$.

Calling the angle $q\,e\,m = \phi$
,, ,, $n\,e\,o = \theta$
,, ,, $w\,e\,o = \theta'$
,, ,, $u\,e\,o = \theta'' = e\,p\,o =$ angle of emergence
,, ,, $w\,n\,e = \sigma$
,, ,, $n\,w\,e = \tau$
,, ,, $\sigma + \tau = \rho = 180° - \phi$

The arc of $\theta - \theta' = \psi$

The line $n\,e = L = f$
,, $w\,e = M = f'$
,, $p\,e = N$
,, $n\,o = x = w\,o$
,, $o\,e = y$
,, $r\,e = z = \dfrac{R}{2}$

R being the common resultant of f, f', f'', in $a\,b$.

Then $L + M : L - M :: \tan \tfrac{1}{2}\phi : \tan \tfrac{1}{2}(\theta - \theta')$

$$\tan \tfrac{1}{2}(\theta - \theta') = \frac{\tan \tfrac{1}{2}\phi\,(L - M)}{L + M.}$$

$\theta = \dfrac{\phi - \psi}{2} \qquad \theta' = \dfrac{\phi + \psi}{2}$

But $\sin \theta : L :: \sin \rho : 2\,Y$
as the the diagonals $s\,e$ and $w\,n$ mutually bisect

and
$$2y = \frac{L \sin \rho}{\sin \theta}$$

but as $L + M : L - M :: \tan \frac{1}{2} \rho : \tan \frac{1}{2} (\sigma - \tau)$

$$\tan \tfrac{1}{2} (\sigma - \tau) = \frac{(L - M) \tan \tfrac{1}{2} \rho}{L + M},$$

and $\not> \tau = \frac{1}{2} \rho$, + arc corresponding to, $\tan \frac{1}{2} (\sigma - \tau)$
$\not> \sigma = \frac{1}{2} \rho$ — the same arc.

Again, $\sin \tau : L :: \sin \phi : 2x$

$2 x = \dfrac{L \sin \phi}{\sin \tau}$ = the distance from one fracture to the other diagonally opposite

$$2y = \text{the resultant of } f \text{ and } f'$$
$$N : y :: 1 : \tan \theta''$$
$$\tan \theta = \frac{y}{N}$$

which gives the angle of emergence, or that made by the polar a, b, of the subabnormal wave, with the horizon.

$$\cos \theta'' : y :: 1 : z$$
$$z = \frac{y}{\cos \theta''}$$

$2 \times \dfrac{y}{\cos \theta''} = R$, the common resultant in the polar $a\ b$,

and, $2 y \tan \theta'' = f''$ the vertical component.

An extremely easy method may be practised of finding the path of a subabnormal wave by an observer in the field.

Referring to Fig. 44. Let a line be stretched across the top of the walls (or anywhere below that, but horizontally), from the exterior or interior angle of fracture, on one wall, to that on the other, w to n, and the length be divided in the proportion of $e\, w$ to $e\, n$; if from the dividing point $'p$, a plumb-line be dropped, it will lie in the vertical plane in which the path of the wave is situate. Let now another line be stretched, or a light straight edge of wood be held, between the points $'p$ and p (corresponding to the line $o\, p$, of Fig. 46); lastly, stretch a line from the point $'p$, so that it shall be square to the line $'p\, p$, and holding it in the hand, "sight it," to coincide visually with the plumb-line: this line or string will then be, in the path of the subabnormal wave, and its azimuth and inclination, may each be at once got, by compass and clinometer, or by two measurements, without the latter instrument. This method admits of quite sufficient accuracy, if the fractured-out pyramid be not too large, but such, that either a straight edge (a straight rafter or joist will answer, of which plenty may generally be found loose about) or a stout cord can be stretched tight across, from w to n. The direction can be thus obtained within a degree or two at most.

It is obvious that if the path of this wave, be referred

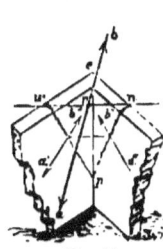

Fig. 41.

to its component path, in either of the two walls, by a plane, normal to a vertical plane, and both passing through the wave-path, then the former plane, will cut the surfaces of the walls, in directions perpendicular to the fracture in each respectively, as in $a'\, b'$ and $a''\, b''$ (Fig. 44), which coincides with what was stated before, as to the general fact, that the lines of fissures from subnormal waves (*i. e.* those emergent in the

THE CATHEDRAL, PATERNO

VALLETTA

plane of the wall) are perpendicular to the path of emergence.

The position of the point p, or the distance down the quoin, at which the two adjacent fractures meet, apart from any variations caused by differences of masonry, &c., depends both upon the velocity, and the angle of emergence of this wave. For referring to Fig. 47, if $a^2\ b^2$, be the path of emergence, referred to one of the walls, and $w^2\ p^2$, be the line of fracture therein corresponding, then as the mass thrown out is greater, as the velocity is so, and as the angle of fracture with the quoin is constant, while the emergence is so, it follows

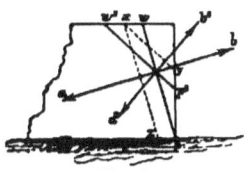

Fig. 47.

that if the velocity be reduced, the fracture will be somewhere as at $x\ y$, still parallel to $w^2\ p^2$, but higher up, and *vice versâ*, and so for any other emergence: but if the velocity being the same, the angle of emergence vary, then, in order that the line of fracture may still continue perpendicular thereto, the point p must ascend or descend along the quoin as with the path $a\ b$, producing the line of fracture $w\ p$, which, when it makes a very acute angle with the quoin, and therefore the angle of emergence small, and the velocity also great, may even follow back along the walls as to $x\ z$, so that the point p, would fall below the base of the wall, if the lines of the fractures were produced.

In the Photog. No. 49, which gives a very good illustration of this class of fracture, as observed at Auletta, this was actually the case. In this view, one of the other large fissures, corresponding to n and k (Fig. 45) may be remarked, as also in Photog. No. 50, at Paterno church.

In the Photog. No. 51 (Coll. Roy. Soc.), also in a street in the town of Polla, a mass, thrown by an extremely oblique subabnormal wave, will be observed to the left hand, in which the effect of this obliquity, and of the miserable class of "nobbly" rubble masonry, and of a floor within, have perturbed the phenomena, as respects the wall, seen most nearly in the plane of the picture. The direction of shock, was in this instance, nearly along the line of the street, towards the spectator, and a little from the left, towards the right; and in further evidence of the general direction here, it should be remarked that the whole side of the street to the right hand where the fronts of the houses are nearly in the line of shock, and *supported as well as held in* by the floors, &c., against the small transverse component of shock, all remains standing, though propped here and there; while at the far end of the street, the fronts of the houses in the street running obliquely across that going from the spectator are all down.

CHAPTER VIII.

SHOCKS OF VERTICAL OR NEARLY VERTICAL EMERGENCE—EFFECTS ON RECTANGULAR BUILDINGS.

WE now arrive at the fifth and last class of waves, viz., those of *vertical* or very *nearly vertical emergence*, upon which it is necessary to make some remarks.

As a strictly normal or abnormal wave, *i. e.* with *perfectly* horizontal transit (unless by reflection) is impossible from a focus beneath the surface, so an absolutely vertical emergence is in strictness limited, to the single point of the earth's surface vertically above the focus, or to the seismic vertical itself. Inasmuch however, as in reality the disturbance producing impulse is not confined to a mathematical point, but (whatever be its nature) extends over a greater or less, and in all cases a very considerable area, and lies also at such a considerable depth, that lines extending from it to the surface make but small angles with each other; so a tolerably large area is found in the midst of every earthquake-shaken region within which the angle of emergence is so steep that it may be viewed, as respects the effects of the wave, as practically vertical.

In the preceding remarks upon the four first orders of wave-paths, we have viewed the wave itself as of one sheet and the movement of any particles in the wave to be forward and backward in a right line or in an elliptic or some other closed curve, in a vertical plane, neglecting the transversal movement (which is small) for the present. This, although perhaps not physically true, was sufficient for our purpose.

Were we at liberty to consider the vibration of the wave with vertical emergence performed in like manner, the movement of any point of matter upon the surface due to it would be limited to two motions—one directly up and down, due to the amplitude, and the other forward and backward horizontally, due to its altitude and to the senses. The movement would thus be similar to that of a normal or subnormal wave. The universal testimony, however, of those who have experienced these vertically-arriving shocks is of a twisting and wriggling motion in different *planes*, violent in its changes of direction, and attended by a movement up and down of *much greater range*, to which the word "sussultatore" is often applied. It seems highly

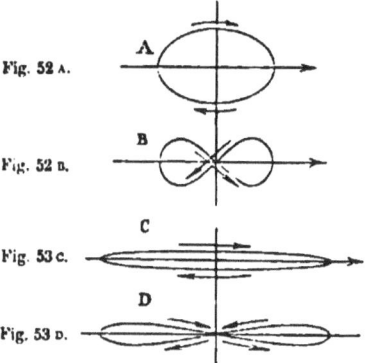

Fig. 52 A.

Fig. 52 B.

Fig. 53 C.

Fig. 53 D.

probable that the path of a wave particle moving normally or nearly so, may be elliptic, as in Fig. 52 A, in a vertical plane, and with a smaller transversal vibration performed in a horizontal plane as in Fig. 52 B. The actual relation of altitude and amplitude, so far as observation yet is afforded, seem more as in Fig. 53 c, or even in a far higher ratio to each other, and the transversal vibration still smaller, as in Fig. 53 D, so that in normal or slightly emergent waves the transversal movement is little noticed, (as, indeed, is true *pro tanto* of the movement in altitude);

hence, as stated, we may neglect for present purposes, the transverse movement altogether as respects such waves.

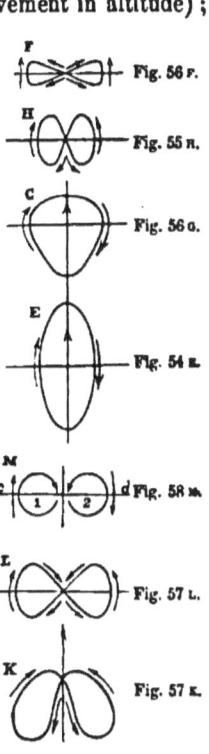

Fig. 56 F.

Fig. 55 H.

Fig. 56 G.

Fig. 54 E.

Fig. 58 M.

Fig. 57 L.

Fig. 57 K.

In the case of vertical emergence, however, the path of a moving particle being in one vertical plane elliptic, and the major axis being the line of transit, will have the form of Fig. 54 E, in a vertical plane at right angles to the former, and in an horizontal plane at the earth's surface the path will be as in Fig. 55 H. In ascending through heterogeneous formations even the more complex forms of Fig. 56 G and F, or Fig. 57 K and L, may be those of the path of the wave particle, the vertical being the movement of largest range, in every instance. In either of those cases the sensible effect upon the earth's surface must be the same as if bodies, besides being lifted up and down, were alternately whirled round in small circles in the directions of the bent arrows 1 and 2 in Fig. 58 M. In vibrating elastic masses having special and pendulous vibrations of their own, when set in motion by the wave, the axis $c\,d$ may rapidly, though with a much slower relative movement, rotate in the direction of the external arrows or in the reverse one. The formidable torsional and wrenching strains which are known to arise from vertical shocks, are most probably thus produced.

It must be remarked, however, that these torsional strains—"Vorticosi" of the Italians and Mexicans—must not be supposed capable of producing those twistings of objects upon their bases, such as vases, chimneys, obelisks, &c., of which we shall record many examples, but which are due to other circumstances first explained by myself several years since.*

A continuous jarring movement, consisting of a rapidly arriving series of waves, moving in a horizontal plane, as in Fig. 57 L, often occurs, and in lofty buildings, such as churches or towers, when the time of torsion vibration of the building itself (once set in motion), happens to be isochronous with that of the wave vibration, twisting strains of enormous violence result.

The effect *upon the walls*, then, of the vertical wave is chiefly to produce fractures which are transverse to the lines of twisting distortion. As the twist is alternately in opposite directions, these fractures cross each other, the opposite contained angles being double, those of the lines of maximum torsion strain with the vertical. These motions being accompanied by rapid up-and-down ones of much

* 'Trans. Roy. Irish Acad.;' vol. xxi. p. 1, 1846.

greater range, each distinct story of the building acquires a separate momentum of its own, in virtue of the weight and attachment to the walls, of the floors and objects upon them. All the fractures tend therefore to separate and close again, as the wave makes its transit; and the several masses, moving horizontally at the same time, with a rotation alternate and increasing as it ascends the building, the replacements do not often coincide with the displacements, and in a few seconds the stability of the walls may be so far destroyed, that the whole falls to the ground in ruin the most complete. Fissures running horizontally or nearly so from the quoins are not unfrequently observable where the emergence is very steep or nearly vertical, examples of which may be observed at both sides of the N.W. wall of the church of Pertosa (Photog. No. 25, page 42).

With vertical or nearly vertical emergence also, gravity acting with inertia, in the first semiphase of the wave, upon the masses of masonry situated directly above doorways, windows, and other such apertures, their tendency to come down is great; and hence, not only are vertical fissures formed over such openings, but they are *open widest at bottom*, one of which will be remarked breaking across the stone lintel over the west door of Pertosa church (Photog. No. 25), but diagonal fissures crossing the piers between windows, where there are doors or other opes beneath, in a lower story, of which examples occur in the Palazzo Palmieri at Polla and will be observed in the Photogs. 178 and 180, Part II.

It is upon the heavy Italian roofs and floors, however, already described, that the most instant and formidable effects are produced by vertical emergence. Upon these the vertical velocity produces a moment of inertia acting directly downwards, and therefore favoured by gravity. Arched roofs, groining, and that form of arched ceiling constructed of hollow pottery, then spread the walls, as they come down, and falling upon the floors below, bring them down in succession. The details of movement will, however, be best given further on.

Upon the whole, the phenomena of vertical emergence, afford little ground for exact observation, with a view to trace the elements of the shock, and their limited occurrence is not to be regretted, on this ground. When once seen they present general features by which they can almost always be recognized with tolerable certainty, but not such as will enable us to ascertain directly, the line which produced downwards, should intersect the centre of impulse beneath the central field. That must be sought for otherwise, by observations at a greater distance from the seismic vertical, where the wave movements have become more uniform, and less complicated. When ascertained by the method of intersections of wave-paths, aided, if occasion serve, by determinations of velocity of transit, applied to the values of the angle e, its correctness may be tested and controlled, within certain limits, by the coincidence or not, of the focus thus obtained, with the *observed area of vertical phenomena*, somewhere within which the seismic vertical is situated.

CHAPTER IX.

CONDITIONS AS TO FORM AND STRUCTURE IN BUILDINGS, WHICH MODIFY THE EFFECTS OF SHOCK—FIVE PRINCIPAL CONDITIONS.

It will now be necessary to refer to the special effects produced, by the architectural figure, and other particular conditions of buildings, &c., upon the principal phenomena due to the shock upon them, as just described. The treating of this might give rise to almost endless details; reference will therefore only be made to a few of the salient conditions, and their modifying effects, from which the intelligence of the observer can deduce, all others that may come before him.

What has preceded has been exemplified, by an imaginary roofless rectangular building, on level ground, and without any apertures. We have now to discuss how the actual conditions of buildings as we find them (having those of Italy in view primarily, however) affect and modify the results. We may discuss this under the following heads of effects, produced by the wave of each class, as modified by the following conditions, viz. :—

    1st. The form, magnitude, height, and unsymmetrical character of construction.

2nd. The form of the surface or foundation, and the relations of buildings in juxtaposition, as in towns.

3rd. The class of masonry and materials, their flexibility and elasticity, &c.

4th. The reactions on roofs and floors, and of these again, upon the other parts of buildings.

5th. The effects of apertures in walls, as gateways, doors, windows, &c.

After these some observations upon the perturbing effects produced by the physical and geological features of the country, will conclude the first part of this Report.

The final dissolution and fall of every compound structure in masonry, occurs by the successive development, of two stages of dislocation. Were it possible that a building could be overturned by the shock, as a whole, the conditions of its overthrow, would of course depend simply upon its "moment of stability." This, however, owing to the imperfect bond at the joints, and the relations of magnitude, to the strength of the materials, can never occur, except in the case of very lofty towers or minarets, spires, or single columns, &c. In such a case the mass may be *overturned*, and dissolution then takes place principally by the stroke of its fall to the ground.

In all other cases, however, the great fractures are produced in the first instance, which break up the whole, into a number of distinct, and more or less independent masses. The immediately subsequent movements of each of these, depends upon its own moment of stability, in as far as it is unsupported, or unaffected by the stability, or by the fall of others. The disintegration of the building—viz., whether it shall split up at all—depends, as we have seen, upon—

    1st. The direction (as to horizontal obliquity and emergence) and the velocity of the wave, directly, and the density of the materials.

    2nd. Upon the tenacity and bond of the materials, inversely, and upon the form and magnitude of the structure.

Both these relations being modified, by the elasticity and flexibility, of the materials.

The *directions of the great fractures* producing disintegration depend upon—

    1st. The direction (as to obliquity and emergence) of the wave; *i. e.*, on the abnormal and emergent angles.

    2nd. The form of the structure as a whole, and of its several parts, or details, previous to severance.

Lastly. The fall of each separate mass, if then severed, depends upon its own moment of stability. If any such separated mass fall, it does so in the direction of its least horizontal dimension: or as for any practical purpose of seismic observation, the moment of inertia due to the oversetting force must lie in this direction (for otherwise the mass falls by twisting), we may limit the consideration to that case.

If t be the thickness of the bed joint of masonry upon which the mass overturns, cut by a vertical plane passing through the centre of gravity of the mass and the line of transit of the wave; i, the slope of the joint, if any, to the horizon; r, the fraction of t, that measures the distance of the point where the line of resistance cuts it, from the mid-length; r', the distance from the bisecting point of t, to where it is intersected by the vertical through the centre of gravity of the mass, W its weight, and Σ its moment of stability; then

$$\Sigma = W \times (r \pm r') \, t \cos i,$$

according as the line of resistance and the vertical through the centre of gravity, are at the same or at opposite sides of the bisection of t.

The weight of the mass (whatever be its form, assumed approaching regular), may be expressed

$$W = \phi \times l\,b\,t \times \delta$$

ϕ being a factor, determined by the angles that its three dimensions l, b, and t make with each other, and on its form, and δ the specific gravity of the masonry. Therefore

$$\Sigma = \phi \, (r \pm r') \cos i \times l\,b\,t^2 \times \delta.$$

If F be the force necessary to fracture the mortar at the joint t, acting in the direction of the wave transit,

$$F + \phi \, (r \pm r') \cos i \times l\,b\,t^2 \times \delta$$

is equal the total resistance of the mass to being overturned by the force of the shock, acting at the centre of gravity in the same line, but opposite direction

$$V \times \phi \times l\,b\,t \times \delta.$$

For similar forms of the fractured and separated masses,

therefore, and like direction of emergence and velocity of wave, the resistance to fracture F, depends upon the cohesion per unit of surface and total area of fracture, and the resistance to fall Σ, upon the density of the masonry, the height, the breadth, and the square of that one of the two horizontal dimensions, which lies in the direction of the shock, and may be called the thickness.

The mass being *severed free*, from all others, by fracture, it depends upon the value of $(r \pm r')$, and upon the emergent angle of the wave, whether it shall fall forward, in the direction of the wave transit, or in the reverse one. And from the nature of the applied force (being that of inertia), δ disappears.

Such are the general conditions as to equilibrium, upon which the fracture and fall of the separated masses, producing final dissolution of the building or structure, depends, and from which, equations for various architectural forms and conditions may be deduced.

The circumstances of fall of simple rectangular buildings have now been explained. Cruciform buildings, such as churches, are affected much in the same ways, the twelve sides of such a building being, in fact, capable of being viewed or arranged, as separable into those as three simple rectangular ones.

Polygonal buildings are rare, and when the number of sides are few, and the length of each considerable, do not present features materially different.

Cylindrical buildings, or conic frustra, however, which may be viewed as polygons of an infinite number of sides, have some peculiarities.

Whatever be the direction of shock, if horizontal, upon

such a building, its effects are the same. The distinction of normal and abnormal wave does not exist as respects them; and unless the angle of emergence of the wave be extreme, or nearly vertical, the lines of fracture are the same in every case, viz., vertically through the axis and transverse to the line of shock. This arises from the fact that the area of fracture, and therefore the total resistance, F, due to it, augment rapidly as the angle of its obliquity with the vertical, through the cylindrical walls increases. So that although the direct tendency of the wave $a\,b$ is to throw off a cylindric ungula, $e\,d\,c$ (Fig. 59), by a fracture

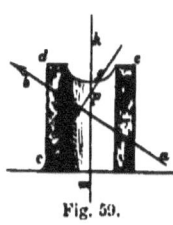

Fig. 59.

perpendicular to its direction in $c\,f$, yet the surface of fracture is so great, and its direction at either side of the vertical plane of the wave transit, so obliquely through the joints, that the building always parts in the weaker line, by diametrical vertical fissures through $k\,m$. The separated masses have now each a moment of stability, the fraction t^2, in which is enormous, being equal to the radius of the cylinder or the base of the cone; and hence the fragments of such towers are seldom overturned.

Where the value of F is small, as in the very bad rubble masonry of the ancient towers, and the angle of emergence considerable, however, we have instances of the mass thrown out assuming the form of a curved ungula, obviously by the fracture commencing vertically, and following down the joints *gradatim* from k to p, and thence to c; of this a remarkable example occurs at Átena.

When the emergence is still more vertical, and the shock powerful, a number of nearly equidistant fissures form round the top of the walls, commencing vertically, and masses are thrown out, each carrying down with it an ungular-shaped toe, so that the tower becomes shorn around the summit at a greater or less distance down, by the descent of several separate masses from it, in the form of Fig. 60, and, as better seen in the Photog. No. 61, page 42, at Marsiconuovo. The original fissures are here, no doubt, produced chiefly by the twisting movements, transferred quite round the walls, by transversal vibrations as referred to when treating of the wave of vertical emergence.

This applies but to a limited extent to the semi-cylindrical "apses" which form the chancel ends, of so many of the more ancient churches. These generally split off at or near the re-entering angles of the quoins, by which they are joined on to the body of the building, if the wave transit approach a line, transverse to that diameter; if otherwise, the walls of the "apse" split vertically in the direction of the opposite diameter.

Fig. 60.

Towers of extreme altitude and very narrow base, such as slender minars, single columns, lofty campaniles, &c., involve a number of complicated and curious considerations, as respects their resistance to fracture and to overthrow by shocks.

In these the elastic modulus, density, and range of flexibility before fracture of the masonry, the time of vibration of the structure viewed as a compound elastic pendulum— together with the direction of wave transit, and the *relation* that may subsist between the amplitude of the wave, its maximum velocity, and the pendulum time and range of vibration of the tower—all are elements. It is not requisite for our present purposes, however, here to pursue the investigation.

The effect generally, of want of symmetry in the severed masses, is to reduce them by further dislocation, prior to complete overthrow. For example: a portion of a uniform wall, severed by transverse fissures from the remainder, but having a buttress of its own or of less height somewhere along its length, is again transversely broken close to the buttress, the moment of resistance to fall being different in each. The relation of the buttress to the wall, as a support against transverse forces of a statical sort, is no longer the same, when the overthrow is produced by a force applied with the rapidity of the wave of shock; there may not be time, to transmit its own stability to the remainder of the wall.

Where the buttress is at the same time a tower rising much beyond the height of the remainder of the building, these generally tend to mutual destruction; the primary fissures occur at the junctions or near them; the walls and the tower have different times of vibration, as elastic pendulums of different lengths, and whether by chance isochronous or not, produce mutual damage, by their impulses upon each other. This is peculiarly striking, in the case of many of the meaner class, of Italian rural churches, where the belfry tower is built into one of the quoins of the main rectangular building, the two adjacent side walls are frequently completely destroyed by transverse rocking of the tower; although the latter may have only suffered fissuring at the lower portions, and that which was above the level of the church walls be overthrown.

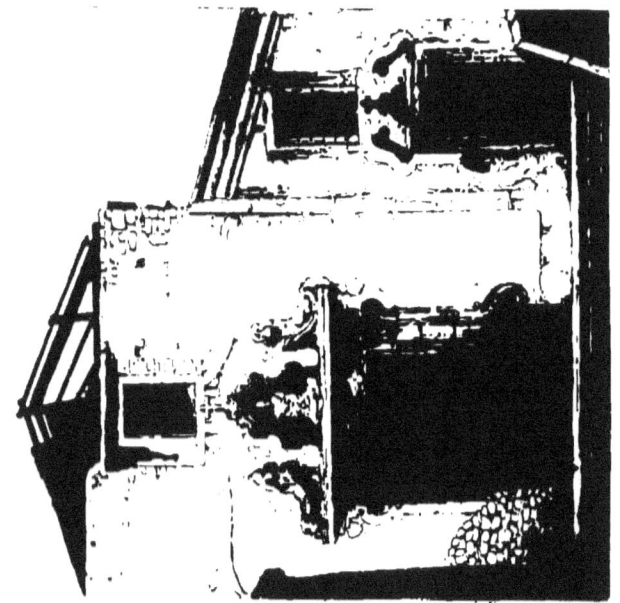

Plate Pl. 82

Sketch. Pl. 82

High towers and of narrow base, fall as one mass, breaking off diagonally somewhere above the base, whatever be the direction of the wave. When, however, the angle of emergence is very steep, a certain amount of shearing force is introduced, and the angle of severance becomes very sharp likewise, so that a sharp angular "aiguille" of shattered masonry remains standing, often bearing a considerable proportion to the whole original height. A remarkable example of this form of fracture is given in Fig. 62 of the tower of the monastery of Santa Dominica, at Montemurro, sketched from the top of the Palazzo Fino, although in this instance probably due only to accidental causes, and not to steep emergence. Isolated fragments thrown from the summits of such towers, owing to their own velocity of pendulous vibration, do not always, by reference to the observed distance of projection, represent without correction, measures of the true direction or velocity of the wave. They are thrown like a stone from a sling, with a certain velocity and direction due to the shock, plus or minus, another, or perhaps a different direction and velocity, due to the proper motion of the tower; of this the observer requires to be on guard.

Unsymmetrical construction of building, always involves unsymmetrical phenomena of dissolution. If compelled to adopt such a building, (in lack of better,) for observation, the first thing to be done to disentangle the phenomena, is to consider the effects, due to the want of symmetry alone. If, for example, we find the opposite walls of a cardinal church, one standing and the other prostrate, the wave transit having been abnormal, and nearly in their lines of length, the first point to be ascertained is, was the prostrate wall symmetrical in form and structure with that remaining. Unless the effects of the roof may have overthrown it, we shall generally find, that the fallen wall was either of much inferior masonry, or smaller thickness, out of plumb originally, or full of windows and doors, the standing wall being solid.

CHAPTER X.

EFFECTS DUE TO FLEXIBILITY AND ELASTICITY OF THE MATERIALS IN BUILDINGS—FLEXIBILITY OF BRICKWORK.

ALL the preceding observations of course have taken no account as yet of the reactions produced on the walls by roofs and floors: they refer to the walls considered as standing alone. The actual extent of elastic flexibility of stone and brick masonry, especially of the former, is not commonly considerable; and unfortunately, as yet, no *precise* measures of these exist for any class of masonry. Were it not for this property, however, no building would stand, even a very moderate shock; and were the velocity of the wave confined within the limits of the velocity of the centre of oscillation of the structure, considered as an elastic compound pendulum, whose time of vibration is due to the length of a simple pendulum equal to the height of that centre above the base, and were the amplitude of the shock within the limit of elastic displacement of the masonry, &c., at that centre, no building would be thrown down.

A well-constructed brick and mortar wall, of 30 or 40 years' induration, and 40 feet in height, unsupported, of two bricks, or 1·60 feet in thickness, has been observed by myself, to vibrate nearly 2 feet transversely at the top, before it fell, in a storm of wind; and that not until after many such oscillations had disintegrated many of the horizontal joints, and produced several vertical fractures. The point of greatest flexion traversed along the length of the wall, as each oblique gust of wind impinged upon it, like the waves of a rope suspended from one end, and jerked transversely at the other.

An octagonal brick chimney stalk, with a heavy granite capping 160 feet in height above the ground, and 15 feet diameter at the base, was observed by me, instrumentally, to vibrate in a moderate gale of wind, when a few months built, nearly 5 inches at the top.

These are illustrations of the extent of flexibility in good brickwork, which possesses it in a far higher degree than stone masonry, the bond of the mortar being better, the flexibility greater, both in the brick and thick mortar joints, these very numerous, and the elasticity more nearly alike in both, than in stone masonry. When the joints are much fewer in proportion, the stone relatively to the mortar, highly elastic and rigid, and the bond, so far as adhesion of the mortar is concerned, small, (indeed, in the case of many hard, siliceous stones, such as granite, almost *nil*,) the result of this difference is, that a well-built and indurated brick wall, when fractured, breaks indifferently nearly, through joints and bricks; but in stone walls, the line of fracture is confined to the mortar joints, with rare exceptions, the rigidity of the several blocks, transferring the whole of the compressions and extensions due to the strains to the mortar alone. From this cause, it was observed very uniformly throughout this earthquake region, that when brick construction was superimposed upon stone

AT POLLA

work, as not unusual in churches, the brick-work, although of so much less density, fell as one mass, with fractures of severance along the lines of junction of the two; and *vice versâ*, when the brick-work, as in a few cases, was beneath, and stone-work above, and when the latter was thrown, if it did not push the brick-work over in its fall, the latter remained comparatively unharmed.

The limit of flexibility of stone masonry exposed to earthquake shocks depends, to an immense extent, upon the flatness and superficial area of the beds of the individual stones, and the completeness with which " breaking joint" and " thorough bonding" are preserved in the setting.

When the masonry consisted of rounded, lumpy, quadrated ovoïds, of soft limestone, as already mentioned in the general description of the poorer and older towns, and of which the Photog. No. 63, of a part of Polla is an example, the whole dislocation occurred through the enormously thick, ill-filled mortar joints; and almost all buildings thus formed, fell together at the first movement, in indistinguishable ruin. In the Photog. No. 64 (Coll. Roy. Soc.) of Pertosa, a poor, but more modern town, the class of masonry was a little better, and, as may be remarked, the ruin less complete.

Where, as in a few examples observed, the masonry was of the best class (and such as would be so recognized in England), the buildings thus constructed, stood absolutely uninjured in the midst of chaotic ruin. Some examples of this will be found in the second part—none more striking than that of the Campanile of Atena, a square tower of about 90 feet in height, and 22 feet square at the base, in which there was not even a fissure, while all around nearly was prostrate. This tower was, however, also aided by iron chain bars, built in at each story. The great viaduct carrying the military road at Campostrina is another example. Indeed, it was evident, that had the towns generally been substantially and well built, or, rather, the materials scientifically put together, very few buildings would have been actually shaken down, even in those localities where the shocks were most violent, and their directions the most destructive. Thus the frightful loss of life and limb, were as much to be attributed to the ignorance and imperfection displayed, in the domestic architecture of the people, as to the unhappy natural condition of their country, as respects earthquake.

In a wall of parallelopipedal blocks, properly overlapping and breaking joint, the aggregate tenacity, of a vertical serrated transverse line of joints, may be represented, as Professor Rankine has shown, by an equation of the form

$$T = n \frac{n+1}{2} f, \delta \times b\,h\,t$$

the last letters being the dimensions of the wall at the serrated section: n, the number of courses; f, the co-efficient of friction, which may equally be taken as the co-efficient of *ad*herence of the mortar, irrespective of its own *co*herence, and d, the specific gravity of the stone or brick. Unfortunately, we still need better experimental data as to the adhesion of mortar, in directions both parallel, and transverse, to the bed surfaces, to enable us to apply the numerical results to earthquake-applied strains producing vertical or inclined serrated fissures. The like difficulties do not arise with horizontal transverse fractures.

The strain is here applied almost with the rapidity of a blow. Almost the whole stress falls instantly upon the less elastic mortar joints, at their surfaces of contact with the stones, when exposed to direct pull, and the mortar joint parts off from the stone with a resistance of only *one-half that due to its statical adherence, or to its statical coherence.* This fact is rendered familiar to the senses, by the facility with which two bricks from an indurated building, that would require a slowly applied load of perhaps half a ton to tear them directly asunder, may be caused to part and drop asunder, by a slight blow from a hammer upon one of the bricks while the other is held in the hand. As applied to our subject, this sufficiently indicates, that the portion of the total force of shock required to produce fissure, or horizontal fracture of the base, of the severed masses, to permit overthrow, is, when these are large, relatively very small; so much so, where the masses, are large in relation to the surfaces of fracture, and the co-efficient f, of adherence very low as in the case of the Neapolitan provincial mortar, that it may be frequently neglected in calculations of seismic statistity. At p. 139, *et seq.*, will be found the method of calculating the velocity due to *fracture of the horizontal mortar joints*, at the base of walls overthrown, which is the most important and frequent case of fracture that occurs, in seismometric observation.

CHAPTER XI.

SECOND CLASS OF CONDITIONS MODIFYING EFFECTS OF SHOCK—FIRST ASPECT OF RUINED TOWNS.

This disposes, by anticipation, of the third class of modifying conditions. As respects the second, nothing is more remarkable and puzzling to an unpractised observer, who enters a town situated upon tolerably level ground, than the apparent caprice by which the fall of the buildings is characterized. He finds one whole side of a street cleared down; turning into the next, within a few hundred feet, he needs be on the watch, to discover any signs of injury. Further on, whole districts of streets have disappeared, and one heap of rubbish, stones, and beams occupies their place; yet, not far off, long lines of houses are but fissured. In some large streets the houses are down here and there, in the most irregular order, some of the very loftiest stand pretty safely—some of the humblest are in dust.

These abrupt changes from safety to destruction are still more remarkable, if the town, in place of resting upon the plain, or on pretty level ground, occupy the summit and flanks of some "colline," or conoidal hill. Here, perhaps, at one side of a line passing over the crest of the hill, nearly everything is demolished; at the other, little damage has been done. Until by the help of such observations as have been described, upon *some one* or more of the *injured but not overthrown large and cardinal* buildings, the observer has got some clue to the direction of transit of shock, he is completely bewildered; and if he has no clear notion formed on this point, he will be much disposed to coincide at first, with the opinions chattered around him by the principal townsmen, the Syndic, the Judice, the Sotto Intendente, &c., who accompany him, and assure him that the shock was in *every* direction " tutti, tutti direzioni," —that it was " orizontale e vorticoso;" and point out, in proof of this, that the buildings have actually fallen in all possible directions, which is undeniable.

A town formed of streets of adjoining and abutting houses, may be viewed, as respects each block of buildings, as a vast cancellated single structure, in which, in certain directions at least, every portion adds stability to all the rest.

The larger the single block of houses, or the longer the line at the sides of a street, if the shock be in the direction

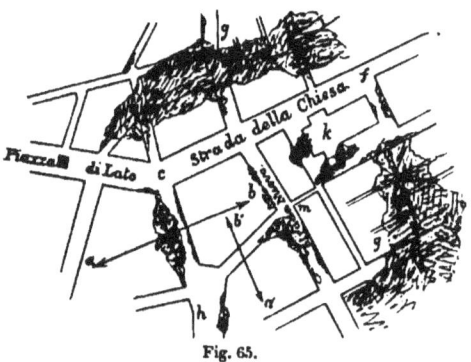

Fig. 65.

of its length, the less, *cæteris paribus*, is the injury done; but there are other causes also in play. Those of some of the principal phenomena may be best illustrated by reference to Fig. 65, an imaginary block plan, of part of an earthquake-shaken town. Whole districts at $q\,q$ are prostrated. At both ends of the great church, and about m, and along the street from h to e, houses have fallen, and others are grievously injured; yet these are perhaps amongst the best built and least ancient houses in the place. Passing along the main central street, after debouching over heaps of stone at e, it is found, that along the greater part of its length, fissures here and there, a few chimneys, or a side wall down, as we look up some "vico," are all the signs of earthquake visible.

The great church, when examined, tells us that the horizontal direction of shock was from a to b. Returning now over the ground, and examining the place with the compass in hand, we find that the main street presented *its length* to the direction of movement; its stiff front and rere walls, its multiplied floors, have saved it. The street e to h, on the contrary, although of houses of the same date, height, and character, &c., is nearly destroyed; it was nearly *transverse* to the line of movement. The street at m has escaped much better, though much injured, yet it is nearly parallel with $e\,h$, and the houses are as high if not higher; but we find they are some of the *newest and best-built* houses in the town. Some houses at the right of the Chiesa Madre are thrown down, but no others in the Piazza, have suffered. The rubbish has probably all been cleared from the Piazza in the market-place, and we cannot tell which way the houses fell, but we look at the fallen apse or campanile or transept, and find that it fell against one of these, and upon these houses. We now get out of the highways of

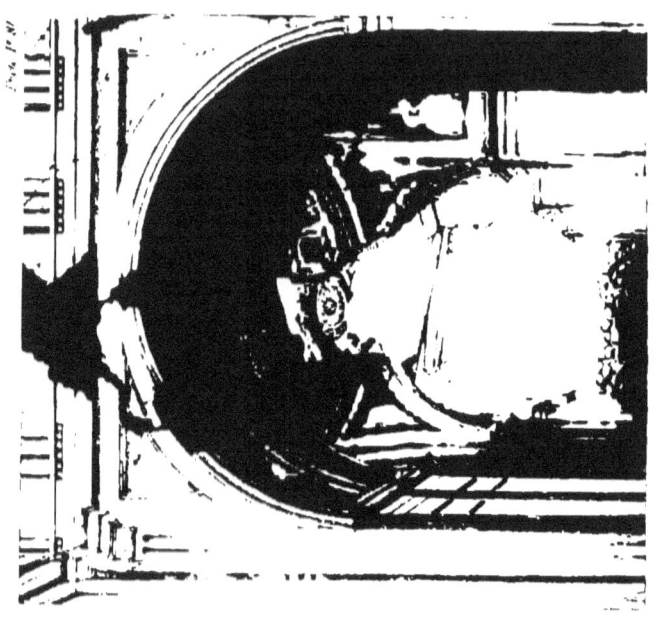

the place, and find a large district whose length is almost parallel with the well-preserved main street and yet is a mass of rubbish. No house stands to tell us of what they were; but, on examining the character of the stones, and of ·the timbers, the shattered doors and window-shutters, and so forth, we find this was one of the poorest quarters of the town, and, as always happens (except in rapidly growing cities), one of the oldest. The buildings here are in heaps, because *before the shock they were tottering* and ready to fall. The whole matter has become clear, and the fact has been learned, that entire towns give little information to the seismic observer, and that but of a general and often uncertain character—that single buildings are his proper objects.

Had the direction of shock been a' to b', in place of what was assumed, a little examination of the figure will make it evident that the whole train of phenomena would have been changed. Hence the prodigious complexity of phenomena presented by towns that have been subjected to two or more shocks in different or in orthogonal directions in quick succession. It is needful to state these circumstances, trivially simple as they appear, because hitherto they have been unnoticed and unregarded by previous earthquake describers. Even the scientific reporters of Calabrian and other earthquakes, who multiply such facts, seem to have had no clear conception of their cause and connection.

The Photogs. Nos. 66, (67 and 68, Coll. Roy. Soc.,) illustrate also some examples. In No. 66, a street in Polla, the direction of shock, was nearly perpendicular to the plane of the picture, and from the observer. The face of the houses, built well and with a batter, here receiving the shock nearly end on, remained in great part standing, and so saved all beyond them, while all lower down that part of the city was destroyed.

In part of a street in the humbler part of Pertosa, Photog. No. 67, (Coll. Roy. Soc.,) the shock was nearly in the perpendicular, to the plane of the picture, and from the observer. One side of the street, that on the right hand, is almost quite down; the opposite one was comparatively safe. The houses were much of the same character at both sides; but on one side the shock *pulled out* the fronts, that have fallen away from the roofs and joists, which had no hold upon them owing to the method of construction, while it *pressed inwards*, against the firmly resisting edges of the floors and roofs the other side, which the wave, in its return vibration, had not sufficient velocity in its horizontal component to bring down.

In Photog. No. 68, (Coll. Roy. Soc.,) some of the highest houses in Polla were left standing tolerably safe, though full of apertures, and arched galleries in front; while the same street, lower down and near the front of the picture, was choked with the ruins of houses, of only half the height or less. The former were *new and well built* in comparison with the latter.

If in place of the town standing upon pretty level ground, as previously assumed, it stand upon a hill summit, and on its flanks, the difference of devastation, at different sides, is usually great. If, as in Fig. 69, the shock emerge in the direction $a-b$, the buildings at the right and left, are differently affected, because their grasp upon the ground at their foundations is different, the relative direction of emergence different (as will be more fully examined when treating of the relations of geological formation to shock), and often (as at Saponara) chiefly because the houses that fall at the steep side to the left hand, fall against and upon,

Fig. 60.

those below them, and the accumulated ruin falls in an avalanche of rubbish down the hill-side.

So also if, in place of a steep angle of emergence, the line of shock were nearly horizontal, as from left to right, in the figure, the whole of the steep side of the town, during the first half vibration, tends to fall *from* the hill-side, and therefore *away* from all the natural abutments that their foundations excavated from the hill-side afford to the sides nearest the hill-top, but all of which generally are operative in *favour* of the houses at the other side of the hill.

This difference may be much compensated by the nature of the rock formation beneath the town, as well as by the form of the hill, &c., but this belongs to a future part of the subject.

CHAPTER XII.

EFFECTS OF SHOCK UPON BUILDINGS—THE 4TH MODIFYING CONDITION—RELATIONS OF FLOORS AND ROOFS.

LET us proceed now to the relations of floors and roofs to the shock, and the effects of both in modifying its action, immediate and final, upon the other parts of the buildings.

The common construction of provincial Neapolitan floors and roofs, has been already briefly described. Were the floors of an earthquake-shaken house perfectly homogeneous, formed, as if of a single parallel plate of cement or beton, self-supported and free from all timbering, the lines of fracture would be almost perfectly regular, and follow in accordance with those of the side and end walls; so that with a normal or subnormal wave, fractures would extend crossways, parallel to the walls, and generally uniting the fissures wherever these were situated in the latter; and with an abnormal or subabnormal wave, the fractures would be diagonal, at right angles to the line of transit of the wave, when viewed in plan upon the floor, and generally uniting the wall fissures situated in the adjacent walls, or diagonally opposite; while with a vertical or very steeply emergent wave, fractures would be found crossing each other in an horizontal plane at right angles, or nearly so, but having any degree of obliquity to the planes of the walls, dependent upon the direction of the line of wave transit, if very steeply emergent, and upon the form and relations, as to horizontal figure and height, of the building—or to the latter conditions only, if the transit were perfectly vertical.

The three cases are illustrated in Figs. 70, 71, and 72. In either of the two latter, large portions would become

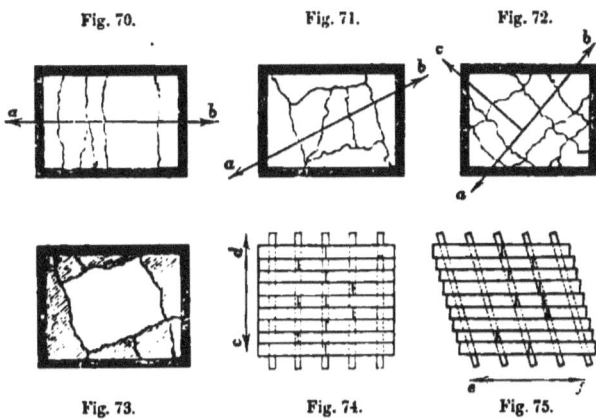

detached, and would fall, leaving other portions, as in Fig. 73, still adherent and supported by the walls; while in the first case, if the width transverse to $a-b$ (Fig. 70) were sufficient, the segments detached by the parallel fractures, would break again transversely, by their own weight at or near the mid-length, and also close to the walls, and then fall.

So far, account has been taken only of direct seismic forces *in the plane of the floors*, but in the subnormal, sub-abnormal, and vertical waves, the inertia of the floor itself is brought into play *transversely* to its own plane, and all the displacements by fall just mentioned as due to gravity alone acting after fracture are produced upon an exaggerated scale by the introduction of this force conspiring with inertia at the moment of shock vertically downwards.

Now these are, in fact, precisely the forms of fracture and destruction, observable in these heavy floors of concrete and tiles, so far as they are left free, in a limited degree, by the constraint of the planking and joists beneath.

The joists and planking, are as one mass, and move together, and parallel to their respective lengths, under forces parallel to either, whether horizontal or slightly emergent. The union, however, is not sufficiently complete, and the jointing of the planking is too rough and open, to prevent "racking" by diagonal forces, as in Figs. 74, 75; and as the several quoins give out unequally, in the case of diagonal wave transit, the buildings are no longer truly rectangular in plan, and so in this the floors follow the walls. If the wave be normal or subnormal, and the *joists* lie parallel to its line of transit, the wall at the end first reached by the wave draws off from the ends of the joists embedded in it. The floor itself, more or less as a whole, follows the wall, and the other ends of the joists draw from the opposite wall, all during the first semi-vibration of the wave. During the second semi-vibration, the whole floor returns by inertia, in the opposite direction to its first movement, and following in the direction of transit of the wave, thrusts back again the previously drawn ends of the joists, into their sockets in the wall *b* (Fig. 76), and the mass is stopped by coming into contact with the interior face of this wall, against which, it *strikes* with enormous violence, the edge of the planking and of the concrete striking the whole length of the wall, almost at the same instant and at the same level. This only happens when the direction of shock is pretty nearly in the line of the *joists*, as a, b (Fig. 76). In such a case, the floor is almost

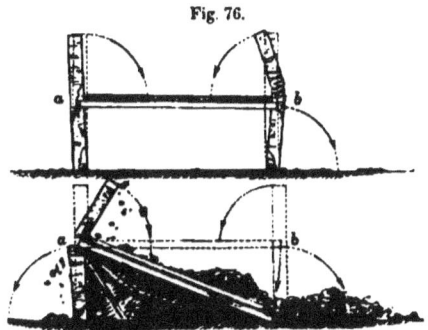

Fig. 76.

Fig. 77.

certain, if of any considerable magnitude, to bring down the end wall, upon which it strikes with the power of a "battering ram."

The sockets of the joists, (inserted, as, has been stated, they commonly are, into the bare masonry without tossal or bond timber,) become partially *occupied* by fragments fallen into them, and on the return into them of the drawn-out ends, these thump out holes, right through the wall, or shake the bond of the masonry effectually. The wall b, for the height of the story below, falls outwards, or in the direction of movement of the floor and wave, but for that portion above the shaken floor, it falls inwards, or upon the floor itself, and, at the moment after the whole support, of the floor beneath at that end has been withdrawn. The end b of the floor therefore sinks and falls, and the other end of the joists, so far as they still remain in their sockets, act as levers, thus loaded, and prize the wall at the end a asunder, so that it falls, likewise, very commonly outwardly, but, by possibility, in the direction towards the floor also. This chain of events, one of the most potent in the destruction of domestic buildings in Italy by earthquakes, is illustrated in Figs. 76, 77.

When there are two or three floors, over each other, that thus carry away the walls, they all (walls and floors) go towards *b*, the walls wholly falling outwardly. It will readily be conceived, what an inextricable mass of confused ruin, houses thus thrown down present, for the side walls that run parallel with the joists (or beams), fissured before transversely to the line of wave transit, are always more or less shaken down also, by the tremendous descent of the floors and walls, upon them falling.

If the line of transit of the wave, be in an orthogonal direction however, parallel to the *planking*, and therefore transverse to the direction of the joists, these having hold of the walls through the intervention of their sockets, the opening of the fissures in the walls on which the joists rest (due to their own inertia), is augmented by the inertia of the floor. The joists, or some of them, in advance of the fissures towards the end *b* in the side walls, going forward with the flooring and side walls, drag the planking from the remainder of the joists, tearing out or bending partially the spikes, or breaking short the trenails; and the main weight of the flooring, thus freed from constraint of the side walls, runs forward as before, and though to a less extent, induces the same train of events, that have already been described.

And when the line of transit of the wave is thus transverse to the direction of the joists, the same set of phenomena result from both the first and the second semi-vibrations of the wave, and with effects proportionate to the velocities in each semiphase. The concrete and tiles of these floors adhere but very slightly to the planking on which they are laid, and the bond is so destroyed by the first slight movement, that the thick and heavy laminum often *slides* whole, or in large fragments, upon the planking, and batters the walls, already inclining outwards, independent of any constraint from the timbering.

When the wave is vertical, or of very steep emergence, the heavy tiled roofing, generally comes down upon the upper floor, almost at the instant that the inertia of each floor, acting at its centre of gravity in the opposite direction to the wave transit, tends to bring it down also. The impulse of the suddenly-imposed load of fallen roofing, conspiring with the effect of the shock, from which the resilience of the joists (if able to have done so at all), have not had time to recover, bring down the upper floor; the united load falls upon those beneath, and the whole are carried away in succession to the ground.

But, circumstances of building and of shock may be such, that the floor does not give way, but that, as an elastic plate or beam, supported and *encastré* at opposite ends (for the planking and concrete, we must bear in mind, have no insertion in the walls, and but slight connection with them), it merely bends downwards by its own inertia and that of whatever be upon it, under the upward stroke of the wave, as by a load suddenly applied over its whole surface. In this case, on the commencement of the second semi-vibration of the wave (or downward stroke), the bended timbers commence to straighten themselves again; their resilience, as a constant force acting through the versed sine of curvature, carries them upwards beyond the line of their passive position, at the same moment, that the second semi-vibration relieves the load upon them, by its inertia acting *against* gravity. The consequence is, the middle parts of a tolerably large floor, spring up with amazing velocity and power, beneath those who may be upon it, furniture is thrown upwards and towards the walls, tiles are projected upwards from the floor surface, masses of the concrete sometimes dislodged, and persons standing or moving on the floor are thrown upwards, and lose their balance.

Such is commonly the source of the strong impression of those who have experienced steeply emergent shocks, of an upward movement, unbalanced by any corresponding downward one, to which the title "sussultatoreo" is usually given, and which, when very violent, is called "sbalza" by the Mexicans. Of the latter, some remarkable examples will be recorded, as narrated to me.

Floors are not always, sources of increased injury however; they may occasionally act the part of props, to walls that if unsupported for their entire height, would have been prostrated, by normal or other waves of the first four classes.

This can only occur, when the line of transit is transverse to the direction of the joists, and when the end wall is in advance of the wave, b, and, consequently, the edges of the floors also, are intersected by a wall whose plane, is parallel with the line of transit, as in Fig. 78; or when some other such fulcrum, resists the forward motion of the floors themselves, and enables them to hold together the side walls by the insertion of the joist's ends, and so to save the wall at cc, dd, by cancellating the structure, in vertical and horizontal directions. The planking connecting the joists, it will be seen, here plays the part of a connecting stay to the side walls cd, cd, at the return stroke or semi-vibration of the wave, tying them together by the inserted ends of the joists, whose distances they themselves fix.

When the floors have given way as described, the building is usually too far

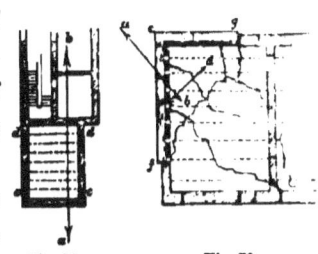

Fig. 78.   Fig. 79.

destroyed, to be of any use for seismometry; beyond this, that a clear comprehension of the *mode* of fall, will always enable the *general direction* of shock to be roughly inferred; but where the floors, being heavy and good, have not wholly fallen, nor the walls, but that these are fissured, and have given out unequally, and the floors also are fissured, but not completely displaced, very valuable indications may be obtained from them, as, for example, in Fig. 79, where the walls $f\,c\,e\,g$, are fissured and given out. We obtain excellent measures of the extent, of this in the directions $f\,c$, $g\,e$, by measurements at the edges of the concrete or tiles, and inside the walls, controlling those of the fissures, which sometimes (though rarely) are not measurable at all, and from the *directions* of the fissures of the floor we may obtain evidence, of another wave movement when occurring in the direction c—d, transverse to the principal one a—b. In fact, the observation of the floors, is only second in importance to that of the walls. The one illustration given, however, must suffice to indicate a large and very varied class of questions to which they may be made to give response.

CHAPTER XIII.

EFFECTS OF SHOCK UPON BUILDINGS—THE 4TH MODIFYING CONDITION CONTINUED — RELATIONS OF ROOFING— MODES OF FALL.

As respects roofing of the ordinary class, of heavy timber framing, with or without common rafters superposed, or with heavy lathing only, to carry the ponderous ridge and furrow tiling already described; all that has been stated of the modes of giving way of flooring applies also, in like cases of wave transit, to it; with this addition, that it frequently happens, owing to the low pitch of Italian roofs, and the enormous weight of the tiling, added to the rudeness of workmanship in the framing, and the want of sufficient iron work in fastenings (iron being a dear commodity, and none made in the kingdom but by the old Catalan process, and all imported heavily loaded with duty); from all this, the roof frames often give way at the tye beams, and the principal rafters then thrust out the side walls, as the roof falls by the shock, if either emergent steeply or vertical. This is almost always the case when the roof is of considerable span, as over churches. In such roofs, "common rafters," are laid upon longitudinal "purlins," or beams stretching across the principals; and when the direction of wave transit is anything nearly horizontal, and in or near the line of the ridge of the roof, the whole roof rocks endways upon its bearings at the level of the eaves or "wall plates," and these purlins act as "battering rams" upon the gable walls, which they almost invariably carry away wholly or in part. A good illustration of this is afforded by Photog. No. 80, of the west end of the church at Picerno, where the common rafters are nearly all gone as well as the tiling, from the right hand side of the roof, and the front gable nearly all thrown down by the E. and W. rocking of the roof, and the inertia of the gables themselves. When the wave is nearly normal, and transverse to the ridge of one of those large heavy timbered roofs, with the ends of the principals resting directly upon the masonry, of the top of the side walls, or occasionally perhaps upon a heavy wall plate of ill-squared timber, the inertia of the whole is so enormous, that in almost every instance, the wall first reached by the wave was thrown outwards, by the shove from the roof and its own inertia together, and the whole roof then dropped nearly plumb down upon, the area of the building, crushing everything before it.

From this tiled roofing, however, whether carried away completely, or only disturbed upon the walls, very few seismometric observations of value, can be made. The chief of these consist, in the occasional indications afforded, by the amount of draw-out of timbers from their sockets in the walls, either longitudinal or transverse. The tiling itself, is so loose, the interstices between the overlaps so great, that it is very seldom *partially* disturbed, never probably *in situ*, unless by nearly vertical emergence of wave, it is either carried away altogether, or presents no signs of movements that can be distinguished, from the ordinary irregularities, looseness, and twistings, of the lines of tiles.

There are two other forms of roofing, viz. arched or groined, and dome roofs, which frequently occur in the better order of churches, and which become highly instructive. Upon these some observations will be necessary. The Neapolitan churches, which are nearly all either Lombardic, Roman, or Palladian, in style of architecture, are very commonly roofed with solid brick (or more rarely stone) arching. The monasteries and public civil buildings, are also very frequently arched, both in floors and roofs, and many of the better built and more modern villas and palazzi, are so likewise, more especially in the upper stories and corridors. Semi-cylindric arching, and hemispherical domes, are the more common forms of church covering of this character, built in brick and mortar, with timbering and tiling over that, or with solid stone sloping pavement, for the outer skin of the roof. Groining of intersecting cylindric arches, is not unusual, in the monasteries and civil buildings, and hemispherical doming, intersecting and connected by cylindric arch-bands, is often met with in the more recent churches. The arching in villa architecture is of comparatively late date, and executed in hollow pottery laid in mortar.

The inertia of all these forms of roofing is very great. Intersecting arched groining universally splits along the crowns of the arches, whenever the direction of wave transit is transverse or oblique, to the line of the axis or springing, and with moderate amount of emergence. If the wave be very abnormal, and the structure pretty large, transverse fissures at right angles or more or less oblique to the former also form. When the emergence is steep, three orders of fissures, if not more, are produced in large arched vaults, one along the crown, and two others parallel to it, and distant from 40° to 50° at either side of it. Where the lateral movement amounts to even a very few inches, the detached masses descend, between those standing at each abutment, enough to destroy equilibrium, and either they fall through, and more or less, from both abutments follows, or the whole comes down together. When the emergence is steep, the two lateral fissures, are further removed from that at the crown, and a very moderate vertical shock, suffices to send outwards, both side walls or abutments, and the whole vault drops between.

The gable ends or semi-tambours of such cylindric vaults, consisting (essentially) of a semicircular plate of masonry resting upon its diameter, level with the vault springing, give out at top at both ends of such vaulted roofing, when the direction of wave transit is along the line of the axis or near it, forming a large fissure transversely, at the junction of the gable and vault ring (or near it), which is usually, most open at the end first reached by the wave, from reasons obvious from what has preceded; the difference is greater as the wave is more subnormal, as gravity conspires with inertia then, to bring out the first reached gable, but acts against inertia, in the second.

When the wave is subnormal and transverse to the axis of a cylindric vault, its greatest overthrowing power is exercised, when the angle of emergence is such, that the line of transit passing through the centre of gravity of the vault, (or of its transverse section), also passes through the joint (either at one side or the other of the crown), that is removed 45° from it, which, as the centre of gravity is about 0·64 r, the mean radius of the arch ring being r, gives an angle of emergence of about 15°. Hemispheric domes, are also most readily, fissured or overthrown, by a wave of small emergence. Where the emergence is steep, the fissures run in curved lines, transverse to the line of transit, and cross each other more or less, when viewed in the line of the vertical axis, and few or none of them pass through the crown. When apertures are formed (such as lights) in the dome, and especially if near its springing the fissures are directed from their upper angles, and great sector-shaped masses are dislocated. An almost vertical or completely vertical shock does not seem to affect domes at all as much, as horizontal or subnormal ones.

The directions of the fissures in any case, depend not merely upon the direction of the wave transit, but also upon the planes of the curving joints, of the structure and upon its details of construction, to such an extent, that general principles can only to a very limited extent be made available for deductive observation from domes.

The Photogs. Nos. 81 (82 Coll. Roy. Soc.), illustrate the general character of the fall of curvilinear roofing. No. 81 is of Tito Cathedral, where the emergence was steep, and the roofing a combination of cylindric vaulting and domes; it shows (see p. 99) the form of a very formidable double fissure in the crown of cylindric arching or vaulting, and the intermediate fragments given down by inertia; beyond this fragments of the lateral domes are visible, and more fully seen in No. 82 (Coll. Roy. Soc.). In No. 168 (see p. 296), the fall of the cylindric vaulting of the nave and chancel at Polla is seen, and its effects in fracturing and sending outwards the side walls at its fall. No. 84 (Coll. Roy. Soc.) is an exterior view of the apse of the church of St. Maria Maggiore at Vignola, where the wave was subabnormal, with moderate obliquity and emergence, and where the fragments of the tower and of parts of its conical roof, were projected on to the roof of the church.

Conical or prismatic roofs of this sort over towers, being generally of timber were not frequently disturbed; when fractured, however, they are so in ways extremely capricious and perplexing. The partial fall of the tiled roof of the apse at Vignola, was due to the drawing away of the heads of the principal rafters from their support by the movement of the curved walls consequent on the large fissures visible in them.

It may be remarked that generally no feature of architectural construction is more characterized by its destructive effects upon the remainder of the edifice when shaken than are those vaulted and domed roofs. Their inertia is enormous, the centre of gravity is high above the walls, and they are deficient in tenacity and flexibility. They therefore not only are dislocated and fractured separately, but their rocking to and fro as a whole on the tops of the walls leads to the destruction of the latter. Upon examining the gigantic ruins at Rome, of the Imperial Baths of Titus, Caracalla and Nero, &c., an eye that has become conversant with seismic observation, at once perceives that the destruction of these enormous edifices, was but little due to the feeble hand of the barbarian, and was mainly produced by the earthquakes that desolated the city between the fifth and ninth centuries, acting thus upon their massive vaults and domes of brick-work.

CHAPTER XIV.

EFFECT OF SHOCK ON BUILDINGS—FIFTH MODIFYING CONDITION—EFFECTS OF APERTURES, ETC. IN WALLS.

We at length come to the fifth and last head of constructive modification, as affecting and affected by, the shock, viz., the effects of wall apertures, windows, doors, &c. A few words of illustration will be sufficient for this. Assuming the simplest case, as in Fig. 86, of a normal

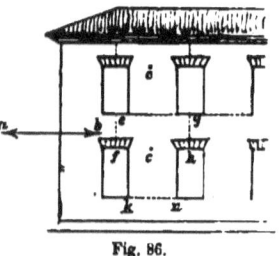

Fig. 86.

wave; whenever the wall is pierced by one or several rectangular or other apertures, it may be viewed as divided into different segments such as that ef, gh, nk, each having a separate moment of stress of its own, and giving rise to a separate dynamic couple, the extremity of one arm being in the centre of gravity, c, the other at the base and junction with the similar adjoining segments, the tendency of the oscillation being, to alter the angular position of each separate mass, and to produce separation from the others in directions parallel to $k\,n$, to $e\,f$, and to $g\,h$.

The fissures tend to form, as in a solid wall, perpendicular to the line of transit of the wave; but as in a solid wall (all of horizontal courses) these must also follow the joints, so must they here where the joints *above* all such rectangular apertures are those of arch voussoirs at various inclinations. Hence, when the wave is normal the fissures form through the nearest vertical arch joint generally, and through to the next aperture above or below, as in Fig. 87; but when subnormal the fractures are through

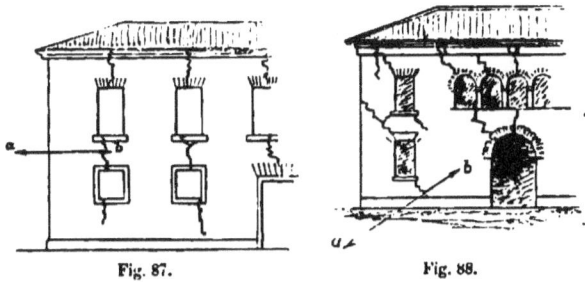

Fig. 87.   Fig. 88.

the joints of the voussoir nearest square to the line of transit, as in Fig. 88; and very generally in such a case the fissures run, from the angles of the apertures. Hence the angle made by the joint, of the plate-band arch or low segment arch, above a window or doorway, through which the fractures run, with the vertical, forms an approximate measure of the angle of emergence. The precise position of the fracture is of course varied by innumerable minuter conditions, such as variation of thickness in the wall, difference of strength at different points otherwise similar, and many others, which must be looked out for, when the phenomena are found perplexed.

Abnormal and subabnormal waves, produce like effects here, to normal and subnormal, except that they are by the former produced in two planes of walls, meeting at an angle usually right, and accompanied by disturbances transverse to the plane of each wall. When the wave is of very steep emergence or vertical, then diagonal fissures are produced in two directions crossing each other, and often accompanied by vertical fissures also, from causes obvious on reference to the statements already made as to the fissures in solid walls due to such waves.

A remarkable example of fissures of this sort will be found in a subsequent page, occurring at Polla. Their general character is that of Fig. 89.

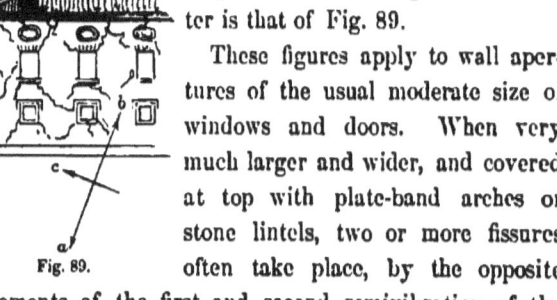

Fig. 89.

These figures apply to wall apertures of the usual moderate size of windows and doors. When very much larger and wider, and covered at top with plate-band arches or stone lintels, two or more fissures often take place, by the opposite movements of the first and second semivibration of the wave, and large fragments fall out.

The usual character of fracture in arches of considerable curvature may be illustrated by the Photog. No. 90 (Coll. Roy. Soc.); but many others will occur in the succeeding pages. When the width of arch aperture is very considerable—eighteen, to twenty or thirty feet for example—and a large mass of wall overhead, fractures transverse to the plane of the wall, usually occur not only at or near the crown of the arch, but (as in the case of roof vaults) 40° to 50° or even more, at either side of it, and the mass above probably descends more or less, and then secondary diagonal fissures are produced by its descent, which have no relation to the angle of emergence of the wave, and must not be confounded, as indicatory of it, with fissures previously produced. An example of this will be recorded in the city of Naples.

Very steep emergence with large semicircular arches, usually produces two sets of fractures also, as in Fig. 91. The wave emergent a to b, produces the fissures $o f$ and $o n$, and those at c and g, probably in its first semi-vibration; in the second semi-vibration the mass $c e f$ tends to rotate round c in a direction a to b and f to e, but the instant it is displaced the weight of the hanging mass $k f n$ breaks the whole across at k by a nearly vertical fissure; $k f$ drops vertically a little, and when $c e k$ has resumed its state of repose, the right-hand side of the soffit of the arch is permanently a little below the left. This might readily be mistaken (alone) for a displacement due to a normal wave.

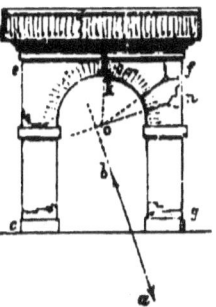

Fig. 91.

Some examples will be given in the narrative also of other singular secondary effects upon the stones and voussoirs of arches by continued oscillation produced by normal or subnormal waves.

When the angle of emergence becomes extremely steep it may occasionally be observed that little or no trace of obliquity of fissure is to be found. The wave, in fact, emerges so nearly vertical that its horizontal elements are those mainly effective in dislocation, and produce only fissures by resolution, where there are piers and window apertures horizontal ones, as in Fig. 92.

Fig. 92.

Stone staircases, the steps of which are bedded into the walls, often produce most complicated effects, both by primary and secondary fracture, and, as objects for deductive information, should usually be avoided.

The choice of buildings best suited for observation in an earthquake region, will have been discerned from what has been stated, though much must always depend upon the observational power and sharpness of the observer, and something upon prior experience (acquirable, however, in a very few days' work). Buildings of the simplest character, large, well-built, not too much injured, and cardinal, are the most important points.

There are many detached objects the observation of which can afford valuable results, in reference to the path of the wave and the direction of its transit, besides the dislocations of the shell of buildings; such as the swing of lamps or candelabra, of hung pictures, &c., and of twisted objects, such as vases, obelisks, &c., which, however, do not demand detailed consideration in this place. Several examples occur in Part II.

CHAPTER XV.

TRANSIT VELOCITY OF THE WAVE FORM.

BEFORE passing to the subject of *velocity of the wave*, a few remarks should be made as to the methods of ascertaining its *transit velocity*. The transit velocity — *that with which the form* of the wave *is transferred* from point to point of the shaken surface — is so great that it can only be ascertained with the desirable precision, by means of a proper seismometer, of the self-registering class described in the author's fourth Report on the Facts of Earthquakes, ('Trans. British Association,' 1858), to be established prior to the shock; and the only known method of determination is based upon observation of the time of arrival of the wave, at each of three or more distant stations, within the shaken area.

In the facts which we can usually collect in the field, *after* the shock, we are limited to the casual observation by the ordinary time measurers (clocks and watches) of the moment of observed shock at several places.

Such observations are liable to multiplied sources of error; from errors in the indicated local time as shown by the time-pieces themselves, and errors of observation of the moment of true shock, as said to have been recorded by them, as well as ambiguous or doubtful statements; so that out of a considerable number of such time-facts, obtained in a seismic-shaken country, probably not above two or three can be really relied upon. Were we in possession of a large number of time records of considerable accuracy, such even as rated chronometers at the distant stations but still more, self-registering instruments, would afford, it would then be important to apply the method of least squares to their discussion, in the way that has been done by Dr. Julius Schmidt, astronomer at Bonn, to the Rhenish earthquake of 1846 ('Das Erdbeben vom 29 July 1846, im Reingebiet,' &c., von Dr. Jacob Nöggerath. Bonn, 1847. Pamphlet, 4to).

If we denote by $\alpha\,\beta\,\gamma\,\ldots\ldots\,\nu$ the surface distances in geographical miles from the seismic vertical $= Z$,

$\mu\,\mu'\,\mu''\,\ldots\ldots\,\mu^n$ their respective differences of longitude in time from Z,

$t\,t'\,t''\,\ldots\ldots\,t^n$ the observed times at A B C .... N,

T, the moment that the wave reaches the surface at the seismic vertical Z,

$\tau\,\tau'\,\tau''\,\ldots\ldots\,\tau^n$ the transit periods or times of running over the distances $\alpha\,\beta\,\gamma\,\ldots\,\nu$,

it is obvious that

$$\tau = t \pm \mu - T$$
$$\tau' = t' \pm \mu' - T$$
and
$$\tau^n = t^n \pm \mu^n - T \quad\ldots\ldots\ldots\ (I.)$$

and that

For places situated to the west of the seismic vertical . . . . . . $\tau = t + \mu - T$ (II.)

And for those to the east of the same $\tau = t - \mu - T$ (III.)
While for those to the north or
south of the seismic vertical or
in the same meridian with it. . $\tau = t - T$ . . . (IV.)
And the respective velocities therefore

$$\frac{a}{\tau}; \quad \frac{\beta}{\tau'}; \quad \frac{\gamma}{\tau''} \ldots \ldots \ldots \frac{\nu}{\tau^n} \ldots \text{(V.)}$$

Obtaining an average transit time per second for three adjacent places, situated in one radius of surface, from the seismic vertical, Dr. Schmidt then applies the method of least squares to the discussion of all the remainder, and with a result undoubtedly important, where, as in his example, he has had thirty distinct observations of time, at as many different stations; but which, when the number of stations is very limited in which any real confidence can be reposed, possesses no advantage over a simple choice from the whole of the most trustworthy, and the reduction from these of the mean.*

The question of transit velocity, however, although of great physical interest, and destined, no doubt, ultimately to connect itself in an important way with that of the velocity of the wave particle (or wave itself), is, as respects seismometry viewed as a branch of physical geology, of subordinate importance at present.

---

* Or, as suggested to me by Dr. Robinson, giving each observation a weight proportional to the length of its own wave-path, which gives the formula—

$$V = \frac{a^2 + \beta^2 + \gamma^2 \&c.}{a\tau + \beta\tau' + \gamma\tau'' \&c.}$$

or, what is the same in result, though perhaps more convenient—

$$V = \frac{a\tau + \beta\tau' + \&c.}{\tau + \tau' + \&c.}$$

CHAPTER XVI.

SECOND CLASS OF DETERMINANTS, OBJECTS OVERTURNED, OR PROJECTED BY SHOCK.

I THEREFORE pass at once to the second class of seismometric determinants—viz., those derivable from the overturning or projection of objects by the shock; *i.e.* by the velocity impressed upon them by the wave itself, and in the direction, of its line of transit, or contrary to it; taking in also, such work of fracture, as may occur in detaching them from their contacts.

And here, as the conditions of *observation* are very simple, being limited chiefly, to the accurate measurements of two ordinates and an azimuth, and to some considerations as to the forms of the bodies, and of their points of attachment, and to the mutual relations of these, we may avoid that prolix detail, which was unavoidable in treating of the deductions, as to direction from fissures and fractures.

CAP. A.—BODIES OVERTHROWN ONLY, *i. e.* WITHOUT FRACTURE OR ADHESION OVERCOME.

I. *By Horizontal Force (Normal Wave).*

From what has been already stated as to the effects of inertia in masses exposed to shock, it is obvious that any loose body, so circumstanced with reference to form and base, as to be overthrown by an earthquake shock (*i. e.* by a simple impulse), may be regarded as a compound pendulum. As the force of the inertia of motion is always $F = MV$, and for the same body proportionate to V simply, so the body may be considered as if struck at its centre of gravity by another body (without loss of *vis vivâ* by impact, &c.) of a weight equal to its own, and moving with an horizontal velocity $= V$. We exclude from investigation, bodies of wholly irregular figure, as such are not fitted for observation, and limit ourselves to such regular forms—prisms, cylinders, pyramids, &c.—as are found in connection with architectural structures or civil life, and are adapted for seismometry.

In order to upset the body, the horizontal velocity impressed by the shock (whatever be the duration of the latter) must be sufficient to make it turn upon one of the arris's or angles of its base, through an angle ϕ formed by the line, Fig. 93, joining the centre of gravity with that angle or arris, and the vertical through it.

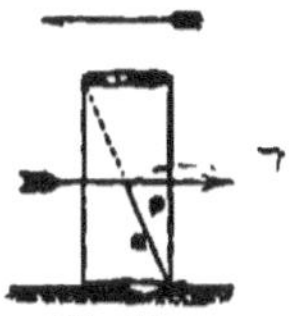

Let a denote the distance (in feet) of this point or edge from the centre of gravity, then the *statical work done* in upsetting the body, whose weight is W

$$W a (1 - \cos \phi).$$

Fig. 93.

This must equal the *dynamical work* acquired, which (as is well known), is equal to the *work stored up in the centre of gyration*, or—

$$W a (1 - \cos \phi) = \frac{W \omega^2 k^2}{2g}$$

where ω is the angular velocity of the body at starting, k the radius of gyration, with respect to the point or arris on which it turns, and g the velocity acquired by a falling body in one second of time.

Equating these two values of the work done we find
$$\omega^2 k^2 = 2 g a (1 - \cos \phi) \quad \ldots \ldots \quad (\text{I.})$$
but ω, the angular velocity, is equal to the statical couple applied, divided by the moment of inertia, or
$$\omega = \frac{V a \cos \phi}{k^2}$$
squaring and substituting
$$V^2 = 2 g \times \frac{k^2}{a} \times \frac{1 - \cos \phi}{\cos^2 \phi};$$
and since the length of the corresponding *simple pendulum* is
$$l = \frac{k^2}{a}$$
$$V^2 = 2 g l \times \frac{1 - \cos \phi}{\cos^2 \phi} \quad \ldots \ldots \quad (\text{II.})$$

In order to apply this formula to any given case we must determine the corresponding value of l, the simple pendulum applying to that case.

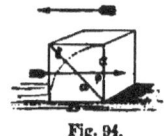

Fig. 94.

1st. In the case of a *solid cube overturned* (Fig. 94) whose side is a;
$$k^2 = \frac{2 a^2}{3}$$
$$a = \frac{a}{\sqrt{2}}$$

therefore
$$l = \frac{k^2}{a} = \frac{2\sqrt{2}\,a}{3}$$
and
$$\cos \phi = \frac{1}{\sqrt{2}}$$
substituting these values in (II.) we find
$$V^2 = \frac{8}{3} g a (\sqrt{2} - 1) \quad \ldots \quad \text{(III.)}$$
and the following geometrical construction is obtained from this expression.

Let $\delta =$ the difference between the diameter and side of the cube $= a (\sqrt{2} - 1)$
$$V^2 = 2g \times \left(\frac{4}{3} \delta\right) \quad \ldots \quad \text{(IV.)}$$

Or, the height due to the horizontal velocity of the wave of shock is equal to four thirds of the difference of the diameter and side of the cube.

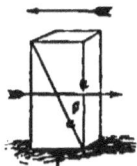

Fig. 95.

2nd. In the case of a *solid rectangular parallelopiped overturned* (Fig. 95).

The altitude being $= a$, side of the base $= \beta$, and $\tan \phi = \frac{\beta}{a}$, we have
$$k^2 = \frac{a^2 + \beta^2}{3}, \qquad a = \tfrac{1}{2} \sqrt{a^2 + \beta^2}$$

therefore
$$l = \frac{k^2}{a} = \tfrac{2}{3}\sqrt{a^2 + \beta^2}$$

substituting in (Eq. II.) we find
$$V^2 = \tfrac{4}{3} g \times \sqrt{a^2 + \beta^2} \times \frac{1 - \cos\phi}{\cos^2\phi} \quad \ldots \text{ (V.)}$$
or,
$$V^2 = \tfrac{4}{3} g \times \frac{a^2 + \beta^2}{a^2} \times (\sqrt{a^2 + \beta^2} - a). \quad \text{(VI.)}$$

Let δ, as before, denote the difference between the diagonal and altitude of the parallelopiped,
$$\delta = \sqrt{a^2 + \beta^2} - a$$
and since $\dfrac{a^2 + \beta^2}{a^2} = \sec^2\phi$

we have the following resulting theorem—

"The height due to the horizontal velocity of wave that will overturn a rectangular parallelopiped is two thirds of the difference of the diagonal and altitude, multiplied by the square, of the ratio of the diagonal to the altitude, or of the secant of the angle ϕ."

3rd. In the case of a *solid right cylinder overturned*.

The height of the cylinder, or altitude, being a, and the diameter, or base, β, as before, we have
$$k^2 = \frac{15\,\beta^2 + 16\,a^2}{48}$$
and
$$a = \tfrac{1}{2}\sqrt{a^2 + \beta^2}$$
therefore
$$l = \frac{k^2}{a} = \frac{15\,\beta^2 + 16\,a^2}{24\sqrt{a^2 + \beta^2}}$$

substituting in (Eq. II.) and since

$$\cos^2 \phi = \frac{a^2}{a^2 + \beta^2}$$

$$V^2 = \frac{15\beta^2 + 16 a^2}{12 a^2} \times g \sqrt{a^2 + \beta^2} (1 - \cos \phi) \quad \text{(VII.)}$$

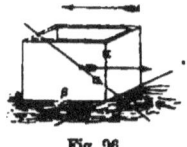

Fig. 96.

4th. In the case of a *hollow rectangular parallelopiped overturned* (Fig. 96).

Let the edges of the parallelopiped be a, β, γ, the thickness of its walls being small in relation to their lengths, and suppose it overturned round the edge or axis γ.

It is easily demonstrable that

$$k^2 = \frac{2\beta(a^2 + \beta^2) + \gamma(2a^2 + 3\beta^2)}{6(\beta + \gamma)}$$

and

$$a = \tfrac{1}{2}\sqrt{a^2 + \beta^2}$$

and therefore

$$\frac{k^2}{a} = \frac{2\beta(a^2 + \beta^2) + \gamma(2a^2 + 3\beta^2)}{3(\beta + \gamma)\sqrt{a^2 + \beta^2}} \quad \text{(VIII.)}$$

from which, substituting the value of l in Eq. II., and remembering that $\tan \phi = \dfrac{\beta}{a}$ we obtain V, the horizontal velocity.

5th. In the case of a *hollow cylinder overturned* (Fig. 97).

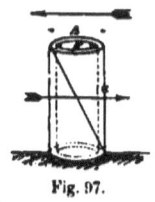

Fig. 97.

Let the altitude of the right circular cylinder be a, the exterior and interior diameters of its base β, β', so that the thickness of the wall $= \dfrac{\beta - \beta'}{2}$

then

$$k^2 = \frac{15\,\beta^2 + 16\,a^2 + 3\,\beta'^2}{48}$$

which is the case of the solid cylinder when $\beta' = 0$, and when the thickness of the wall is so small in relation to the other dimensions that it may be neglected

$$k^2 = \frac{9\,\beta^2 + 8\,a^2}{24}$$

In either case, the distance of the centre of gravity from the axis or point of rotation at the base is

$$a = \tfrac{1}{2}\sqrt{a^2 + \beta^2}$$

Hence, in the first case,

$$\frac{k^2}{a} = \frac{15\,\beta^2 + 16\,a^2 + 3\,\beta'^2}{24\sqrt{a^2 + \beta^2}} \quad \ldots \text{(IX.)}$$

and in the second case,

$$\frac{k^2}{a} = l = \frac{9\,\beta^2 + 8\,a^2}{12\sqrt{a^2 + \beta^2}} \quad \ldots \ldots \text{(X.)}$$

and

$$V^2 = 2\,g \times \frac{9\,\beta^2 + 8\,a^2}{12\sqrt{a^2 + \beta^2}} \times \frac{1 - \cos\phi}{\cos^2\phi}.$$

WEDGES SEVERED OFF.

6th. In the case of a *solid parallelopiped with two adherent wedges, overturned* round the free or external arris.

This form is that frequently occurring, as thrown from the ends of rectangular buildings, and described in treating of fissures, the end wall being the parallelopiped, and the wedges the adherent portions of masonry, fractured from the side walls.

The general and exact treatment of this case involves expressions too complex for practical use. The case should never be appealed to for deciding the value of V, unless the magnitude of the parallelopiped be large, in proportion to that of the wedges, the lower angle of the latter small, and not very unequal at the two ends of the parallelopiped, and the thickness of the walls β, small in proportion to the height a. In that case a sufficiently near approximation is readily made.

Referring to Fig. 98, let the mass of the two wedges

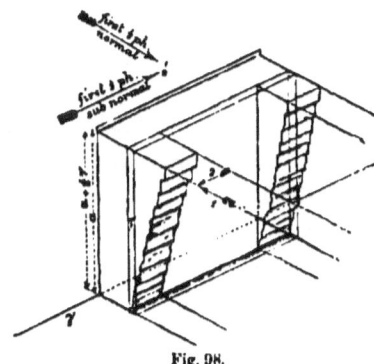

Fig. 98.

be determined, and reduced to a parallelopiped, two of whose sides shall be equal to a and γ, these being the height and length of the parallelopiped, overturning on γ.

Let the thickness or third dimension of this rectangular plate be τ, and let it be supposed applied to that side of the parallelopiped, to which the wedges were adherent, and added to the thickness t, or width of base of the parallelopiped, for the value of $\beta = t + \tau$. Further, h being the height of the end wall or parallelopiped, let its altitude be assumed increased in the proportion

$$t : h :: t + \tfrac{1}{4}\tau : a$$

for the new value of a. The case now resolves itself into that of Eqs. V. and VI., substituting in these the new values of a and β, thus obtained; in any case worth practical application this may be done without sensible error.

Case 7th. *Angular wedge thrown over upon its apex.*

This is the case referred to pp. 66–72, in treating of fissures, as one of frequent occurrence, and valuable in deciding direction of wave-path. It can, however, be very nearly applied to the determination of velocity. The problem, generally treated, leads to very complex results; and approximations are equally tedious, except in the case in which the direction of the wave-path is parallel to one of the external sides of the wedge, when the wedge vanishes and the case becomes identical with the last one.

What has preceded refers only to horizontal force or velocity V (normal wave). We now proceed to

II. Oblique Force (*Submormal, Abnormal, and Subabnormal Waves*).

Let O'O be the wave-path passing through any building whatever, as Fig. 99.

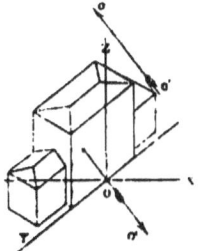

Fig. 99.

Let O X be perpendicular to the lines O Z and O Y, and let a, β, and γ be the angles made by O'O, with O X .. O Y .. O Z.

If V = the total velocity, or that in the path of the wave, and v_x, v_y, v_z the components along X Y and Z, then

$$v_x = V \cos a$$
$$v_y = V \cos \beta \quad \ldots \ldots \ldots \text{(XI.)}$$
$$v_z = V \cos \gamma$$

The effect of v_x in overturning the structure has been already considered. The component v_y produces no direct effect in overturning, although its action parallel to Y may fracture and disintegrate the building.

If the structure is capable of being overturned in the plane of Y Z, and also in the plane of X Z, the components v_x and v_y must act together, and compel it to turn round upon one of the extreme points in the line O Y. In that case, the motion ceases to be comparable with that of a compound pendulum, and is reduced to the movement of a body round a fixed point under the influence of gravity; one which, even in its simplest cases, would be too difficult of reduction for practical purposes, and which virtually never occurs in seismometric observation. If, therefore, the force of the wave be confined to the plane X Z, and be emergent at an angle e with the horizon, we have

$$v_x = V \cos e$$
$$v_z = V \sin e \ \ . \ \ . \ \ . \ \ (XII.)$$

It being remarked that when the wave is not *strictly* subnormal, we may always view it as such in the first instance, and resolve the value of V found, through the abnormal angle, the latter being less than sufficient for longitudinal disintegration of the wall or structure.

The general case of the subnormal wave must be distinguished into two. First, when the wave-path through the centre of gravity falls *within* the base, second, when it falls *without* the base of the structure.

Let O' o' C (Fig. 100) be the wave-path emergent within the base. The structure is urged by inertia against

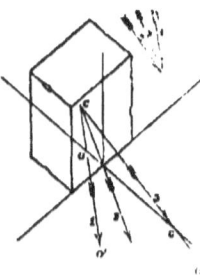

Fig. 100.

the ground in the direction C O' (contrary to the wave in its first semiphase), and it cannot be overturned by that movement of the wave at any velocity, (that is, by the direct shock); but it may be overturned by the wave in its second semiphase (or by the return shock). The limit of this, is where the wave-path, passes through the centre of gravity and the arris of the base, round which the structure should turn, when it is still capable of being overturned by the first semiphase of the wave.

In either case, overthrow by the second semiphase, although mechanically possible, seldom occurs (for the same object), in parts of buildings, as some support or obstacle generally props it against the latter.

Let a body cylindrical or prismatic be overturned by a wave in the path O C (Fig. 101). Inertia of motion, due to

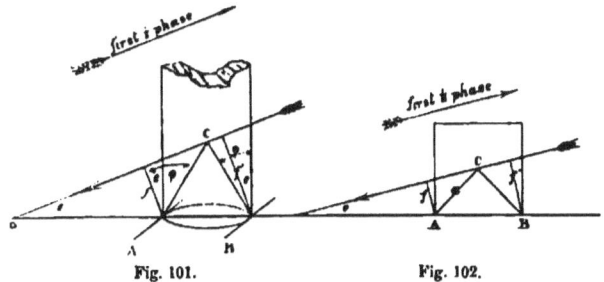

Fig. 101.   Fig. 102.

the *first semiphase*, acting in the contrary direction C O, tends to make the structure turn round the axis A.

The overturning couple is

$$\frac{v_s a - v_s \beta}{2},$$

where a and β are double the co-ordinates of the centre of gravity, or, M being the mass,

$$\frac{MV (a \cos e - \beta \sin e)}{2}$$

Dividing this by the moment of inertia, we obtain

$$\omega = \frac{V (a \cos e - \beta \sin e)}{2 k^2}$$

Substituting this expression in Eq. I., we find

$$V^2 = \frac{2 g a l^2}{\left(\frac{a \cos e - \beta \sin e}{2}\right)^2} \times (1 - \cos \phi)$$

but

$$\tfrac{1}{2}(a \cos e - \beta \sin e) = f = a \cos(\phi + e)$$

or

$$V^2 = 2 g l \times \frac{1 - \cos \phi}{\cos^2(\phi + e)} \quad \ldots \text{(XIII.)}$$

Which, when $e = 0$, reduces itself to Eq. II. for the normal wave, or $V^2 = V^2$.

But if the structure be *overturned by the second semiphase*, inertia acts in the direction of transit O C, and tends to make it overturn round the axis B.

The overturning couple is

$$\frac{M V (a \cos e + \beta \sin e)}{2}$$

and

$$f' = \frac{a \cos e + \beta \sin e}{2} = a \cos(\phi - e)$$

f in the former, and f' in the latter cases, denoting the perpendiculars let fall upon the wave-path from the axes A and B respectively.

Substituting in Eq. I., as before, we find

$$V^2 = 2 g l \times \frac{1 - \cos \phi}{\cos^2(\phi - e)} \quad \ldots \text{(XIV.)}$$

When $e = 0$ or the wave normal, this also reduces to Eq. II., as must necessarily result from the fact that both semiphases of the wave (assuming the velocity practically the same in both) are equally effective in producing overthrow by horizontal shock; the structure presenting *similar aspects* to the wave-path, in both semiphases.

When $\phi + e = 90°$, the wave-path passes through the centre of gravity and axis of overturning, and V becomes infinite, so that the structure cannot be overturned by any velocity of shock during the first semiphase of the wave.

When $e = 90°$, and $\phi = 0$, the wave-path is vertical, and the structure cannot be overthrown in the wave-path by any velocity, but may be conceived *lifted*, in the second semiphase of the wave, by its own inertia of motion first impressed.

And when $\phi - e = 0$, the wave-path is perpendicular to the diagonal, or $f' = a$, and the wave in its second semiphase, produces its maximum effect, that maximum in the first semiphase, of course occurring when the wave is normal.

Proceeding to the consideration of the special problems:—

>8th. In the case of a *solid cube overturned by subnormal wave*.

Preserving the foregoing notation, the structure shall overturn round A (Fig. 102), in the first semiphase, or round B in the second semiphase, of the wave.

a being the side of the cube,

$$l = \frac{\sqrt{2}}{3} a$$

and since $\phi = 45°$ (Eq. XIII. and XIV.) become

$$V^2 = \tfrac{1}{3} g a \times \frac{\sqrt{2} - 1}{\cos^2(45° \pm e)} \quad . . (XV.)$$

the sign + applying to the first and − to the second semiphase of the wave.

9th. In the case of a *solid parallelopiped overturned* (*subnormal wave*).

Here
$$l = \tfrac{2}{3}\sqrt{a^2 + \beta^2}$$

therefore
$$V^2 = \tfrac{2}{3} g \times \sqrt{a^2 + \beta^2} \times \frac{1 - \cos\phi}{\cos^2(\phi \pm e)} \quad \text{(XVI.)}$$

the signs + and − being attended to as before.*

10th. In the case of a *solid right cylinder overturned* (*subnormal wave*).

In this case
$$l = \frac{15\,\beta^2 + 16\,a^2}{24\sqrt{a^2 + \beta^2}}$$

and
$$V^2 = \frac{g}{12} \times \frac{15\,\beta^2 + 16\,a^2}{\sqrt{a^2 + \beta^2}} \times \frac{1 - \cos\phi}{\cos^2(\phi \pm e)} \quad \text{(XVII.)}$$

+ and − applying as before.

11th. In the case of a *hollow parallelopiped overturned* (*subnormal wave*).

Here, from Eq. VIII., XIII., and XIV., we have

$$V^2 = \frac{2g}{3} \times \frac{2\beta(a^2 + \beta^2) + \gamma(2a^2 + 3\beta^2)}{(\beta + \gamma)\sqrt{a^2 + \beta^2}} \times \frac{1 - \cos\phi}{\cos^2(\phi \pm e)} \quad \text{(XVIII.)}$$

---

* Eq. XVI. has been applied in the text of Part II. under the form
$$V^2 = \tfrac{2}{3} g \times \frac{(a^2 + \beta^2)^{\frac{3}{2}}}{(a \cos e \mp \beta \sin e)^2} \times (1 - \cos\phi)$$

12th. In the case of a *hollow right cylinder overturned* (*subnormal wave*).

Here, from what has preceded, we have

$$V^2 = \frac{g}{12} \times \frac{15\beta^2 + 16a^2 + 3\beta^2}{\sqrt{a^2 + \beta^2}} \times \frac{1 - \cos\phi}{\cos^2(\phi \pm e)} \cdot \text{(XIX.)}$$

and

$$V^2 = \frac{g}{6} \times \frac{9\beta^2 + 8a^2}{\sqrt{a^2 + \beta^2}} \times \frac{1 - \cos\phi}{\cos^2(\phi \pm e)} \cdot \cdot \text{(XX.)}$$

Proceeding now to

Cap. B.—Bodies or Structures Fractured.

If the fracturing force, or M V, in the direction of the wave-path, M being the mass broken off, act transversely to the plane of fracture, the case is one simply of cohesion destroyed by an *impulsive* force, in which 2 M V is equal to the *statical strain* that would have produced the same fracture; and if the direction of the force be such as to produce rotation in the mass fractured off, there will be a dynamic couple to be taken into account; and lastly, if the plane of fracture occur so, that it is not transverse to the line of force, the latter may be resolved into one that shall be so, which is all that need be said as to *direct fractures*, such as those passing down vertically or diagonally, as fissures through walls, &c.; and the rather because, precious as these become as *indices of direction*, they should never be adopted as measures of wave-velocity, from the uncertainty that must always attend the knowledge of the coefficient of force necessary to produce fracture *through the joints across* the beds of masonry, &c.

Proceeding, therefore, to the determination of *fracture occurring at the base, or in horizontal planes, or in those of the continuous beds of the masonry*, or through homogeneous bodies, such as stone shafts of columns, &c.—to none of which the same uncertainty of coefficient applies—

First. Let the *wave-path be normal* (the force horizontal).

If any prismatic or cylindrical (Fig. 103) solid structure be broken off, by an horizontal fracture at its base, from its own material below that base, and by a normal wave, neither turning over, nor being displaced, but *tending* to overturn, upon the axis of A, by the first semiphase, and upon that of B, by the second semiphase of the wave.

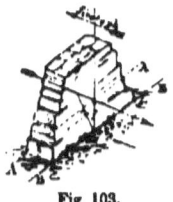
Fig. 103.

The condition for its fracture thus, without overthrow, is that the overturning moment, shall be equal to the moment of cohesion of the fractured surface of the base.

The fracturing force may be considered as applied at the centre of gravity of the mass detached; and the moment of cohesion at half the radius of oscillation of the plane of fracture, at the base, viewed as surface *about* to vibrate round the axis A or B, as a compound pendulum.*

---

* It has been remarked that "this involves the assumptions, (1) that the body will begin to revolve as if it were *absolutely rigid*, and (2) that the force of adhesion, on any element of the plane of fracture, will vary, *cæteris paribus*, as its distance from the axis A, as if the force were not impulsive, but the mass had *extensibility*; and it is asked, is there any experimental law which sanctions this conclusion for impulsive as well as continuous forces? If the mass has *extensibility* in its elements perpendicular to the plane of severance, it must, in like manner, have *compressibility*; and in such case the mass will not tend to turn round the axis through A, but round some axis parallel to it, on one side of which, there will be compression, and on the other

In accordance with the theory of Leibnitz, we therefore have

$$M\,Vf = F\,A\,\frac{k^2}{\beta};$$

M = being the mass of the detached portion;
V = the velocity of the wave in its path (normal),
f = the perpendicular height of the centre of gravity above the base of fracture;
F = the coefficient of dynamic cohesion, or the force upon the unit of surface of the material fractured, which, when suddenly applied, is sufficient to produce fracture;
A = the area of the base of fracture in such units;
k = the radius of gyration of the plane of fracture with respect to the axis A or B.
β = the width of A B.

If W = the weight of the mass broken off, g = the velocity due to gravity in one second, then

$$V = g\,\frac{F\,A}{W} \times \frac{k^2}{f\beta};$$

and if the detached mass be any regular prism or right

---

extension. If the compressions and extensions follow the same law, and have the same coefficient, this axis, or *neutral line*, will divide the base into two equal parts, which would entirely change the necessary amount of the fracturing force." I should admit the correctness of the conclusion thus expressed, if I could altogether, the premises, and their applicability to the matter in hand may be disposed of in a few words. The *actual* extensibility of all building materials, and still more their compressibility, are so extremely small, that for our present purposes both may be regarded as in the text, without sensible error, the compressibility for the small intensity of pressures we are dealing with is insensible, and therefore the position of the axis of rotation is practically that assigned to it. If the objector will point out any good ground for adopting *any other* axis of rotation, I shall be ready to employ it.

cylinder, whose height above the horizontal plane of fracture is a, then $f = \dfrac{a}{2}$. If L be the *modulus of dynamic cohesion*—a coefficient representing the length of a prism of the same material, whose weight is equal to the force upon the unit of surface, if suddenly applied, which is sufficient to tear it directly asunder,—then

$$V = g \times \frac{L}{a} \times \frac{k^2}{f\beta} \quad \ldots \ldots \ldots \text{(XXI.)}$$

or

$$V = 2g \times \frac{Lk^2}{a^2 \beta} \quad \ldots \ldots \ldots \text{(XXII.)}$$

1st. In the case of a *solid cube, fractured from its horizontal base.*

Here

$$k^2 = \frac{a^2}{3} = \frac{\beta^2}{3};$$

therefore

$$\left. \begin{array}{l} V = \tfrac{2}{3} g \times \dfrac{L}{a}, \\[1em] V = 21\cdot 46 \, \dfrac{L}{a} \end{array} \right\} \ldots \ldots \ldots \text{(XXIII.)}$$

or

2nd. In the case of a *solid parallelopiped, fractured from its horizontal base.*

Here

$$k^2 = \frac{\beta^2}{3};$$

therefore

$$\left. \begin{array}{l} V = \dfrac{2g}{3} \times \dfrac{L\beta}{a^2}, \\[1em] V = 21.46 \, \dfrac{L\beta}{a^2} \end{array} \right\} \ldots \ldots \ldots \text{(XXIV.)}$$

or

3rd. In the case of a *right circular solid cylinder, fractured from its horizontal base.*

Here $\beta = D$, the diameter of the cylinder,

$$k^2 = \frac{5}{16}\beta^2;$$

therefore

or

$$\left.\begin{array}{c} V = \tfrac{3}{8} g \times \dfrac{L\beta}{a^2}, \\[1em] V = 20\cdot 12 \, \dfrac{LD}{a^2} \end{array}\right\} \quad \ldots\ldots\ldots \text{(XXV.)}$$

4th. In the case of a *hollow parallelopiped, fractured from its horizontal base.*

Assuming the thickness of the walls small, as before, and γ being that side of the parallelopiped, which is the axis of inceptive rotation,

$$k^2 = \frac{\beta^2(\beta + 2\gamma)}{4(\beta + \gamma)};$$

therefore

or

$$\left.\begin{array}{c} V = \tfrac{1}{2} g \times \dfrac{L\beta(\beta + 2\gamma)}{a^2(\beta + \gamma)}, \\[1em] V = 16\cdot 10 \times \dfrac{L\beta(\beta + 2\gamma)}{a^2(\beta + \gamma)} \end{array}\right\} \ldots \text{(XXVI.)}$$

5th. In the case of a *hollow square prismatic tower, fractured from its base.*

Then, $\beta = \gamma$, and (Eq. XXVI.) becomes

or

$$\left.\begin{array}{c} V = \tfrac{3}{4} g \times \dfrac{L\beta}{a^2}, \\[1em] V = 24\cdot 15 \times \dfrac{L\beta}{a^2} \end{array}\right\} \ldots\ldots \text{(XXVII.)}$$

6th. In the case of a *hollow right cylinder, fractured from its base.*

Here
$$k^2 = \frac{5\beta^2 + \beta'^2}{16},$$

and
$$\left. \begin{array}{l} V = \dfrac{g}{8} \times \dfrac{(5\beta^2 + \beta'^2) L}{\beta a'}, \\[1em] \text{or} \\[0.5em] V = 4{\cdot}02 \times \dfrac{(5 D^2 + D'^2) L}{D a^2} \end{array} \right\} \quad \ldots \ldots \text{(XXVIII.)}$$

In the 5th and 6th cases, and generally in cases of solid columns or minarets, &c., or hollow prismatic or cylindrical towers, the fracture at the base, is never perfectly, and all through, horizontal. When the breadth or diameter is small, however, in proportion to the height, the irregularity of the fracture is not great, and the slope from horizontal also small; and no serious error is introduced by considering the plane of fracture as horizontal.

Secondly. THE WAVE-PATH SUBNORMAL (*force emergent and oblique to the horizon*).

When the wave-path is subnormal, in any prismatic structure, the first and second semiphases of the wave, act upon it as already explained; the former to produce inceptive overturn upon the axis (arris of the base) A, and the latter upon the axis B.

Both cases are included in Eq. XXI.

$$V = g \times \frac{L}{a} \times \frac{k^2}{f \beta}$$

Writing for $f =$ the perpendicular height of the centre of

gravity above the base of fracture, as in the preceding

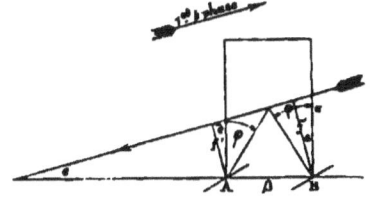

Fig. 104.

equations, f or f' of Fig. 104, and Equations XII., XIII., &c., &c., in which
$$f = a \cos (\phi + e)$$
$$f' = a \cos (\phi - e)$$
and a, as before, the distance of the centre of gravity from the axes of inceptive rotation, we have

$$V = g \times \frac{L\,k^2}{\beta\,a\,a} \times \sec(\phi \pm e) \quad \ldots \ldots \quad (XXIX.)$$

The sign + applying to the first semiphase, and that − to the second semiphase of the wave.

7th. In the case of a *solid cube, fractured from its horizontal base, by subnormal wave.*

Substituting for k^2 its value, and also for a,
$$k^2 = \frac{a^2}{3} = \frac{\beta^2}{3}$$
$$a = \frac{a}{\sqrt{2}}$$

we obtain

$$\left. \begin{array}{l} V = \dfrac{\sqrt{2}}{3} g \times \dfrac{L}{a} \times \sec(45^\circ \pm e) \\ \text{or} \\ V = 15{\cdot}17 \times \dfrac{L}{a} \times \sec(45^\circ \pm e) \end{array} \right\} \quad \ldots (XXX.)$$

8th. In the case of a *solid parallelopiped, fractured from its horizontal base, by subnormal wave.*

Here
$$k^2 = \frac{\beta^2}{3}$$
$$a = \tfrac{1}{3} a \sec \phi$$

and substituting in Eq. XXIX., we find

$$\left. \begin{array}{l} V = \tfrac{2}{3} g \times \dfrac{L \beta}{a^2} \times \dfrac{\cos \phi}{\cos (\phi \pm e)} \\[2mm] \text{or} \quad V = 21\cdot 46 \, \dfrac{L \beta}{a^2} \times \dfrac{\cos \phi}{\cos (\phi \pm e)} \end{array} \right\} \ldots (XXXI.)$$

9th. In the case of a *right circular solid cylinder, fractured from its horizontal base, by subnormal wave.*

Here
$$k^2 = \frac{5}{16} \beta^2$$
$$a = \tfrac{1}{2} a \sec \phi$$

therefore

$$\left. \begin{array}{l} V = \tfrac{4}{5} g \times \dfrac{L \beta}{a^2} \times \dfrac{\cos \phi}{\cos (\phi \pm e)} \\[2mm] \text{or} \quad V = 20\cdot 12 \, \dfrac{L \beta}{a^2} \times \dfrac{\cos \phi}{\cos (\phi \pm e)} \end{array} \right\} \ldots (XXXII.)$$

10th. In the case of a *hollow parallelopiped, fractured from its horizontal base, by subnormal wave.*

Substituting for k^2 and a their values

$$k^2 = \frac{\beta^2 (\beta + 2\gamma)}{4 (\beta + \gamma)}$$
$$a = \tfrac{1}{2} a \sec \phi,$$

we have

$$V = \tfrac{1}{2} g \times \frac{L \beta (\beta + 2\gamma)}{a^2 (\beta + \gamma)} \times \frac{\cos \phi}{\cos (\phi \pm e)}$$

or

$$V = 16\cdot 10 \times \frac{L \beta (\beta + 2\gamma)}{a^2 (\beta + \gamma)} \times \frac{\cos \phi}{\cos (\phi \pm e)} \qquad \text{(XXXIII.)}$$

11th. In the case of a *hollow square prismatic tower, fractured from its horizontal base, by subnormal wave.*

As before, $\beta = \gamma$, and

$$V = \tfrac{3}{4} g \times \frac{L \beta}{a^2} \times \frac{\cos \phi}{\cos (\phi \pm e)}$$

or

$$V = 24\cdot 15 \times \frac{L \beta}{a^2} \times \frac{\cos \phi}{\cos (\phi \pm e)} \qquad \text{(XXXIV.)}$$

12th. In the case of a *hollow right cylinder, fractured from its horizontal base, by subnormal wave.*

Here

$$k^2 = \frac{5\beta^2 + \beta'^2}{16}$$

$$a = \tfrac{1}{2} a \sec \phi;$$

therefore

$$V = \tfrac{1}{8} g \times \frac{L (5\beta^2 + \beta'^2)}{a^2 \beta} \times \frac{\cos \phi}{\cos (\phi \pm e)}$$

or

$$V = 4\cdot 02 \times \frac{L (5 D^2 + D'^2)}{a^2 D} \times \frac{\cos \phi}{\cos (\phi \pm e)} \qquad \text{(XXXV.)}$$

Thirdly. Let the structure be *fractured, from its horizontal base, and also overturned,* whether by a *normal* or a *subnormal* wave.

If the structure be observed fractured only, at the base,

but not overthrown, the velocity impressed was sensibly no more than sufficient for fracture; if it be overthrown also, it was sufficient for both. Hence, if $v_f =$ the velocity determined by the Eq. XXI. to XXXV. for fracture only, and $v_t =$ that for overturning only, by the Eq. I. to XX., the total velocity of the wave will be found

$$V = v_f + v_t \quad \ldots \ldots \ldots \text{(XXXVI.)}$$

It may occur, that a structure shall be fractured from its base, but not overturned, (merely caused to oscillate within narrow limits), by the first semiphase of the wave; and being so broken, may be overturned in the direction of wave-transit, by the second semiphase; in such an example (which is of rare occurrence) the change of sign, in the second members of the equation, must be attended to, and also whether the proper velocity of the mass, viewed as a pendulum, in returning back upon its base, may have conspired with the velocity of the wave itself, in its second semiphase, to overturn the body. In such an example, if the wave be subnormal, with a pretty large angle (e), the impressed velocity will generally be found sufficient, to have projected the structure (if falling entire) to some distance from its base, as well as to have overturned it.

CHAPTER XVII.

VALUES OF THE COEFFICIENT L.

BEFORE concluding this section, it remains to assign the values of the coefficient L for practical use.

It consists of two factors: the tenacity or resistance to rupture, by a force suddenly applied; and the specific gravity of the mass fractured off, by direct pull from an unit of section.

When a direct force, producing fracture by extension, is gradually applied to any prism, whose length and section are both unity, the work necessary to produce the rupture is

$$W = \tfrac{1}{2} P l \quad \dots \dots \dots \dots (\mathfrak{A}.)$$

P being the static load gradually applied, and l the amount of extension of the body on the unit of length at the limit of rupture. But if P be applied at once (suddenly), then $2W = Pl$, the accumulated work, is twice that necessary for fracture, or $\dfrac{P}{2} =$ the force, whose *tension suddenly applied*, as by an earthquake shock, shall rupture the prism.

This force we suppose applied by the weight of a prism of the material fractured, whose base is the unit of section fractured; or δ being the specific gravity

$$L = L \delta = \frac{P}{2} \quad \dots \dots \dots (\mathfrak{B})$$

When the question relates to the fracture of a homogeneous body—such as a column shaft, of one block of stone for example—then the force P to be taken, is that which applies to the material, and δ its sp. gr. But when the fracture occurs in walls of whatever sort, it takes place by the giving way, by loss of adhesion (generally), or sometimes of its own cohesion, of the mortar or other cement, as being the weakest part of the heterogeneous mass: in which case, P is to be taken for the rupturing force of either the adhesion or cohesion (as the case may be) of the mortar or cement, and δ the specific gravity due to the *whole mass* of masonry.

Fracture seldom or never occurs through the solid stone, in masonry, but always at the mortar joints, and generally by their loss of adhesion, to the stone at the faces of the joint. It rarely occurs through the brick, in brickwork, and only when the cohesion of the brick itself, is less than that of the cement.

To enable these equations to be applied generally, in earthquake countries, I have arranged the two following tables, I. and II., which embrace almost all the reliable information we as yet have, applicable to the matter, and from which, the value of L may be deduced, for a great variety of cases.

Many of the numbers, for want of better experimental data, can only be viewed as approximative.

The most important numbers by far, are those relating to the adhesion and cohesion, of the varieties of common mortar; and, fortunately, these have been ascertained by Boistard, Gauthey, Treussart, and Colonel Totten, with considerable accuracy.

The use of the coefficient L, in Eq. XXI. *et seq.*, considers the value of l (Eq. ℨ) evanescent, so that the prism at the moment of fracture has not risen through an appreciable angle, at the surface of fracture, and from the extremely small extensibility of mortar, stones, &c., this is sufficiently true to nature.

TABLE I.

Factors for the coefficient L.

MATERIAL.	1. Weight in pounds per cub. foot. Sp. gr.	2. Pounds per square inch. Resistance to Pressure.	3. Pounds per square inch. Resistance to Tension.	4. Authority for 1 and 2.
Limestone, Caserta, Naples	170	8173	908	Rondelet
Upper Limestone, Genoa	169	4917	546	Gauthey
Jurassic Limestone, Givry	148	4232	496	..
Cretaceous Limestone (Compeigne)	154	3007	334	Rondelet
Lava, Hard Vesuvian	166	8735	972	..
Lava, Soft Vesuvian	107	2209	246	..
Lava, Piperno (Pozzuoli)	162	8140	905	..
Travertino, Old Roman	147	2297	255	..
Travertino, Pæstum	141	3102	345	..
Peperino, Roman	123	3135	347	..
Tufa, Old Roman	78	797	89	..
Tufa, Naples	82	718	80	..
Hard brick	98	1851	206	..
Soft ill-burnt brick	91	1200	133	..
Mortar, lime, and sand, unground	102	423	47	..
Ditto, ditto, ground	119	577	64	..
Mortar, Pozzolano, of Rome and Naples, unground	92	503	56	..
Ditto, ditto, ground	105	732	81	..
Mortar, Old Roman (Campagna)	97	1047	105	..
Mortar, Old French (Bastile)	94	753	84	..
Plaster of Paris (mean)	..	500	55	Laisne

It appears, from the few experiments that have been made, that the resistance of stones, &c., to tension, varies from $\frac{1}{8}$th to $\frac{1}{10}$th the resistance of the same material to compression. The third column is here calculated on the mean of such data. It cannot be viewed as more than an approximation, except in the cases of mortars, which are from actual experiment, as given by Gauthey ('Sur la Construction des Ponts'), and by Rondelet ('L'Art de Bâtir').

TABLE II.

Of the specific gravities, cohesion, and mutual adhesion, of various building materials. Factors for the coefficient L.

MATERIAL.	Weight per cub. foot. Specific gravity.	Resistance to Tension. lbs. per square inch.	Adherent Resistance. lbs. per square inch	Authority.
Granite	164	1200 ?	..	T.
Granite to Portland cement	..	..	97	W.
Granite to Parker's cement	..	..	22	W.
Silurian slate	170	2300 ?	..	T.
Oolite (Portland)	132	270	..	W.
Oolite to Portland cement	..	..	146	W.
Oolite to Parker's cement	..	..	42	W.
Sandstone, coal measure	147	234 to 250	..	T.
Millstone grit and Portland cement	..	..	76	W.
Sandstone (Whitby) and Portland cement	..	..	57	W.
Kentish rag and Parker's cement	..	..	29	W.
Brick, best English	135	200 to 230	..	..
Brick, inferior	97	40 to 80	..	B.
English brick in Portland cement	107	..	..	W.
Portland cement	127	400	..	W.
Parker's cement	120	300	..	W.
Mortar (sand 3, lime 1)	100 to 119	11 to 20	0.88	B. G.
Green and fresh	..	2	..	T. T.
Mortar, ground lime and tiles	100 to 120	40 to 80	5.26	B. G. / T. T.
Hydraulic mortar	..	..	..	..
Jurassic limestone to mortar	..	..	3.80	R.
Brick and tile to mortar	..	..	8.27	R.

Authorities.— T., Tredgold. W., White, 'Trans. Inst. C. E.' B., Barlow. B. G., Boistard and Gauthey, 'Const. des Ponts.' T. T., Treussart and Totten. R., Rondelet.

In the preceding table, the cements had, in all cases, six months to indurate, and the mortars (except in the second case) from six months' to seventeen months' induration.

Examples of very old and good, lime and sand mortar, may be found occasionally in good brickwork—such as that of the Roman amphitheatre at Pozzuoli, for example; or in rubble masonry, where the bond of the stone with lime mortar is peculiarly strong, as with the oolitic building stones, and limestones generally, and with a few sandstones and porous traps, in which the adhesion, of the indurated mortar, becomes fully equal to its cohesion, and both rise above 50 lbs. to the square inch, for forces suddenly applied.

In determining the mean specific gravity, of brickwork and rubble masonry, the proportion of mortar, to the brick or stone in a given volume, may be taken at from $\frac{1}{12}$th to $\frac{1}{4}$th, according to the goodness of the work.

TABLE III.

Deduced values, under different conditions, for the coefficient L.

No.	Conditions of Fracture.	Value of L.
1	Apennine limestone, broken through the stone	225
2	Cretaceous limestone, ditto ditto	154
3	Apennine limestone rubble masonry, of best quality, broken through the joints..	52
4	Apennine limestone rubble masonry, of inferior quality, broken through the joints	30
5	Apennine limestone, rubble masonry of best quality, mortar not indurated..	3·9
6	Argillaceous rubble masonry of the Murgia (Apennine marl rocks), best quality, with indurated mortar ..	55
7	Best Italian or Roman brickwork in mortar	63
8	Inferior brickwork in mortar	30
9	Brickwork, the mortar not yet indurated	2·5
10	Rubble masonry of tufa and mortar, good, with mortar indurated	87
11	Rubble masonry of Travertino, or Peperino, and mortar indurated	51

These values of L, are all, for the mortar when yielding *in cohesion*. When observed to yield *in adhesion*, the coefficient in each case becomes 0·1,L for brickwork and 0·083,L for limestone. The values given, are also all for ancient and fully indurated (except 5 and 9) mortar; where the latter is under twenty-five years laid, the value of L should be taken (*quam prox.*) at ⅝ L in the table.

Proceeding now to

CHAPTER XVIII.

FOMURLÆ REFERRING TO CAP. D.—BODIES OR STRUCTURES PROJECTED.

LET a body, A (Fig. 105), such as the coping-stone of a wall, a church bell, a ball or finial, upon a tower summit, &c.,

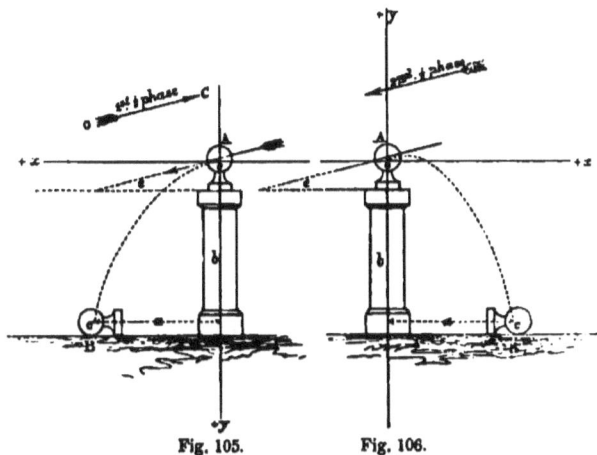

Fig. 105.   Fig. 106.

be thrown from its place by the earthquake wave in its first semiphase (direct shock), in the direction of wave-path O C, and be found projected to the ground at B, in a direction contrary to the wave transit. It is required, if the angle of emergence e, of the wave-path at the place, be known, to determine the velocity of projection, and *vice versâ*. The body is projected downwards throughout its trajectory; motion is imparted to it, in virtue of the grasp that its base had of it, by adhesion or otherwise; and the velocity of projection impressed, or that which, of the total velocity of the wave at its moment of maximum, is effective in projection, is the difference, between the maximum velocity of the wave, and that which is destroyed by adhesion, or other equivalent resistances. The larger the mass the greater is the proportion of the total velocity effective.

Were there no adhesion or equivalent resistance (as in the case of a ball balanced on the top of a staff), the body would drop plumb or nearly so, and might be struck by the base (the wall in Fig. 2) in the second semiphase of the wave; or if the velocity of the wave were infinite or extremely great in relation to g, the body might, whether adherent or not, be displaced and replaced, without projection. These, however, do not occur. The relation in nature between V and g is such, that bodies are projected from buildings, &c., in both semiphases of the wave, and the adhesion of the base is most generally of such a nature as to impart a movement of rotation to the body thrown, which is sufficient to turn it over, more or less (usually from the forms found either through 90° or 180°), during its descent, notwithstanding its high vertical velocity downwards.

Let the axis of y (Fig. 105), be measured downwards vertically, and that of x horizontally from the origin, at the centre of gravity of the body projected, the trajectory described is

$$y = x \tan e + \frac{x^2}{4 \text{ H} \cos^2 e} \quad \cdots \cdots \text{ (XXXVII.)}$$

H, being the height, due to the velocity of projection.

If b denote the height through which the centre of gravity has descended, to reach the ground, or the horizontal plane passing through the centre of gravity, when so deposited, and a the horizontal distance, traversed by the same centre, on striking the ground, then

$$b = a \tan e + \frac{a^2}{4 H \cos^2 e}$$

from which the following expressions are easily deduced for the angle of emergence (which is alternate and equal, to the angle of projection) and for the velocity:

$$\text{Tan } e = \frac{-2H \pm \sqrt{4H(H+b) - a^2}}{a} \quad \text{(XXXVIII.)}$$

$$V^2 = \frac{a^2 g}{2 \cos^2 e (b - a \tan e)} \quad \cdots \quad \text{(XXXIX.)}$$

In the second semiphase of the wave (or return shock) the displaced body is thrown, not downwards, but more or less upwards, if projected by the inertia of motion, acquired from a subnormal wave. If the wave were perfectly normal, the projection of course would be horizontal, and $e = o$ for both semiphases.

In the case of projection by subnormal wave, observing that the axis of y is measured vertically upwards, and that of x horizontally, from the origin, in the centre of gravity of the body as before, we have for the trajectory (Fig. 106)

$$y = x \tan e . - \frac{x^2}{4 H \cos^2 e} \quad \cdots \quad \text{(XL.)}$$

and substituting in this as before

$$y = -b$$
$$x = +a$$

we find
$$-b = a \tan e - \frac{a^2}{4 \operatorname{H} \cos^2 e}$$
whence the angle of emergence or of projection is
$$\operatorname{Tan} e = \frac{2 \operatorname{H} \pm \sqrt{4 \operatorname{H} (\operatorname{H} + b) - a^2}}{a}. \qquad \text{(XLI.)}$$
and the velocity of projection
$$V^2 = \frac{a^2 g}{2 \cos^2 e (b + a \tan e)} \quad \ldots \ldots \text{(XLII.)}$$

As the velocity of projection by earthquake-shock has been proved, by the examination of this shock of December, 1857, to be small, and therefore H, the height due to it, also small; we can find either V or e, geometrically, by the application of Prof. Galbraith's very beautiful problem, for determining graphically, either of these quantities for a projectile; and as this method may be applied by any unmathematical observer, who measures on the ground, the vertical and horizontal heights of a body thrown, and can use a pair of compasses, it will be well to transcribe it.

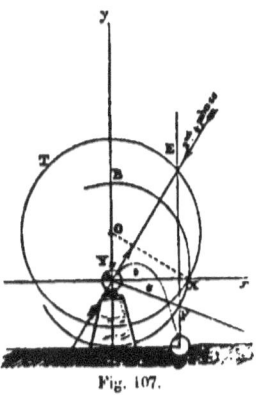

Fig. 107.

Let A (Fig. 107) be the top of any tower or other elevation from which a body has been projected. From A draw A B vertical and = 4 H (H being the height due to the velocity, supposed given). Through A draw A X horizontal. Bisect C B in Y, and on B C describe the semicircle B X C. Bisect B A in O, and with O as centre and O X as radius describe the circle X T. This is the locus circle (*i.e.* that in which the line of aim—in our case the wave-path, shall cut the vertical drawn through the point D, at which the projectile falls—whether the angle *e*, be above or below, the horizontal line through the point of projection).

Let D be the point at which the projectile falls to the ground; draw D F E vertical through its centre of gravity. The directions A E and A F, formed by its intersections with the vertical, give the superior, and inferior angles of elevation, for the given horizontal range and elevation, and coincide in result with Eq. XXXVIII. and XLI.

The wave-path must always, be either horizontal or emergent. Hence in the first semiphase of the wave, although the motion of the projectile is *contrary* to that of the wave transit, the angle *e*, given by the above construction, will be the *superior* one, and also in the second semiphase of the wave, in which the motion of the projectile is in the *same* direction with the wave transit, the angle *e* will be still the superior one.

The values of V. given by Eq. XXXIX. and XLII. are those of the projectile itself, but are less than the maximum velocity of the earth-wave by the velocity destroyed by adhesion, &c. The latter produces rotation in the body, and we generally find it *overturned*, as *well as projected*. The velocity, therefore, destroyed by adhesion is equal to that which has produced the rotation, *v*, and may be arrived at by the Eq. I. to XX. inclusive, and that velocity so found reduced to the direction of the wave-path, and added to the velocity of projection, will give the total velocity, or

$V =$ the maximum velocity of the wave.

If in the same locality, we are enabled to observe two different bodies, both projected, and to measure the vertical and horizontal distances to the point of fall, we can determine *both the angle of emergence* of the wave-path (e) and the *maximum velocity* of the wave. Thus, for example, let both the bodies, be projected by the second semiphase of the wave, and let $a\,b$ and $a'\,b'$ denote the co-ordinates in x and y, of the two trajectories; then by Eq. XL. we have

$$-b = a \tan e - \frac{a^2}{4\,H\,\cos^2 e}$$

$$-b' = a' \tan e - \frac{a'^2}{4\,H\,\cos^2 e}$$

from which we find

$$\tan e = \frac{a^2 b' - a'^2 b}{a\,a'\,(a' - a)} \quad \ldots \ldots \ldots \text{(XLIII.)}$$

$$H \cos^2 e = \frac{a\,a'\,(a' - a)}{4\,(a\,b' - a'\,b)} \quad \ldots \ldots \ldots \text{(XLIV.)}$$

and substituting for H its value $\dfrac{V^2}{2\,g}$ we find

$$V^2 = g \times \frac{a\,a'\,(a' - a)}{2 \cos^2 e\,(a\,b' - a'\,b)} \quad \ldots \ldots \text{(XLV.)}$$

In the case, of *the upper portion of a wall, thrown off from the lower which remains standing*, which is a very frequent one, the equations to apply, are the same as for a body, projected and overturned from the summit; the upper portion turning over first, upon one arris, and then being thrown more or less from the base of the wall, in a trajectory.

The preceding equations embrace, probably, every case likely to occur to observation.

CHAPTER XIX.

THE PHYSICAL AND GEOLOGICAL FEATURES OF THE COUNTRY SHAKEN.

BEFORE proceeding to Part II. some remarks are required upon the general physical and geological features of the earthquake region of December, 1857, in order that the references as to their modifying effects, upon the directions, local variation of intensity, reflection, &c. of the shock, to be made in Part III. may be understood.

The notion commonly formed, from our books of geography and maps, of the physical configuration of the surface of Italy, is that of a long strip of land, separating into two at the south, and divided right down the midst of each strip, by the ridge of the Apennines, with a steep watershed to either shore.

This is but a very inadequate representation of the facts, and only to a limited extent true. Confining ourselves to the kingdom of Naples—*i.e.*, starting on the north, with a line reaching from the mouth of the Tronto on the east, to that of the Tiber on the western shore; from Monte Pennino (in Roman territory) down through the summits of Monte Corno and the Majella, to near Monte Acuto, south of Melfi—the highest ridges of the southern Apennine chain, are found following a wavy line, at about one-third the breadth of the peninsula, from its northern coast. Again, from Monte St. Angelo, just above Amalfi, to a point approaching the Adriatic coast, some miles south-east of Bari, a transverse ridge stretches nearly west to east and from sea to sea, and which bending southward, to the north of Taranto, continues with decreasing development, down into the extremity of Otranto.

Returning to Monte Acuto, the great central ridge is continued, in a direction almost due north and south, for nearly 150 miles, and then stretches in a waving line, down to the southern end of the Calabrian peninsula, where it culminates in Cocuzzo and Aspramonte.

At the north-western end of the first ridge, we have Monte Corno, nearly as high as Etna in Sicily, with several summits, between that and Acuto, of from 7,000 to 9,000 feet in height. In the transverse ridge, Acuto is the highest crest, probably; but Monte St. Angelo, in the little peninsula of Cape Campanello, terminates the western end, as a rampart to the Bay of Naples, at an elevation of 4,770 feet, the elevation gradually declining from Monte Acuto to the Adriatic.

Again, between Monte Acuto and Capo del Armi, at the toe of Calabria, we have Cocuzzo, 5,620 feet, and Aspramonte, variously stated at from 5,830 to 4,380 feet. The little peninsula of Gargano forms a small mountain system of its own, an elevated well-studded table land, of a lumpy, roundish form, with radiating stream channels, in which Monte Calvo is said to be the highest point, reaching 5,088 feet.

These ridges, in lines far from straight, and broken by many differences of elevation, are, indeed, the spine and ribs of Southern Italy; but our notions thus limited, convey no true idea of the physical features of the country. These ridges determine, the great forms and directions of the water-sheds, but by no means those of the vast tracts of subordinate mountain ranges and culminations, by which these axial chains are surrounded and buttressed.

From the Tronto, to Gargano, the lateral mountains tend on the whole, to stretch parallel to the lines of the rivers, which fall with north-eastern courses into the Adriatic; and hence, the lines of mountain and valley, are generally transverse to the axial chain on this side.

On the opposite side, between the great axial chain and the transverse axis, from Naples to Monte Acuto, the great rivers, such as the Carigliano and the Volturno, take in tributaries from every point of the compass, and indicate, the extreme irregularity that prevails, in the alignment of the secondary ridges. This is also, to a less extent, true of the great trapezoidal area, between Salerno and Monte Acuto on the north, the gulfs of Salerno and Policastro on the west and south, and the southern continuation of the axial chain from Monte Acuto on the east. The southern branches, however, of the largest river within this boundary, the Salaris or Sele, have a nearly south to north course. Eastward of the axial chain of Monte Acuto, and over the whole province of Basilicata, the rivers all run, nearly parallel to each other, and in a direction almost exactly from N.W. to S.E. into the Gulf of Taranto; but the directions of the secondary ridges are, on the whole, distinctly transverse to the river courses, which make their way through breaks or depressions, or wind round the terminals of the short and abrupt ridges. So that on the whole the mountainous country south of the transverse axis, and down to the parallel of 40°, at Policastro, may be viewed largely, as a surface furrowed in parallel ridges, running north and south with a trend westward, but twisted, broken through by gaps, and irregular in a high degree.

South of parallel 40°, to the extremity of Calabria Ultra, the lateral chains, tend to place themselves at right angles to the axial chain, except about the boundary separating the two Calabrias, between Cape Suvero, on the west, and Capes Alice and Colonne on the east coasts, where a mountain knot is formed by the intersection of a short but well-defined transverse axis, running east and west, and nearly parallel with the great transverse axis to the northward; the road over which at Petrania, between Cosenza and Nicastro, reaches an elevation of nearly 3,400 feet.

This transverse chain is, in fact, the dam, that absorbs the earthquake movements of Calabria Ultra, and prevents their full spread northwards, and *vice versâ*, just as the great transverse chain of Monte Acuto, partially stops the propagation northward, of the shocks from the Principatas and Basilicata. At the intersection of the transverse axis of Monte Acuto, with the north and south one, there is a great mountain knot, comprised between Laviano and Oppido, east and west, and Venosa and Potenza, north and south. Within this space, which presents some of the grandest scenery of the Apennines, Muro and Bella occupy almost the central position. The mountainous country thus described, extends over about one half of the entire surface, of the kingdom of Naples. The remainder consists of vast plains, (relatively at least) of two distinct sorts—one, the

rolling, rounded, hilly country, constituting the vast grazing downs, of Capitanata and Basilicata, on which countless sheep and goats are pastured in winter and spring; the other the rich corn plains, level as the sea almost, of which the largest are in Otranto and Bari; next to which come those of the Terra di Lavoro, the plain of Pæstum, and of Calabria Ultra, from Rosarno to St. Euphemia (the scene of the great earthquake of 1784). All these great plains (piani) are on the seaboard, but almost every mountain valley of any magnitude, consists of a piano, almost perfectly level, from the sides of which, the mountains spring abruptly, as from a sea shore. The largest of these, is the Piano di Diano, in Principato Citeriore, the scene of some of the worst disasters of the earthquake of December, 1857, in early spring presenting, as do all these valley plains, characters of the richest and most enchanting country. The general aspect of the Val di Diano may be gathered from the Photogs. Nos. 108 (and 109 Coll. Roy. Soc.), being views of the town of Diano, from which the Vallone and Piano take their name; the other of St. Arsenio, on the west side of the same plain. The smaller valley bottoms, present the same characters upon a less scale—many are partially in forest. The mountain cincture of the piani, usually consists, of one or several sloping terraces of small elevation, having frequently the character, more or less perfect, of "parallel roads," tracing round the margins. Those in the piano of the Bay of St. Euphemia, have been described by Meissonier, 'Comptes Rendus for 1858,' and such terraces are observable around a large portion of the Val di Diano.

The piani are not always, or necessarily, strictly *level*

plains, however; some slope, gently but continuously in one direction.

Of the geology of the kingdom of Naples very little is accurately known. Within the parallels of 40° to 42°, the following are the leading facts so far as I have observed them, and learned from the sketch map of Italian geology of Colligno. Probably the lowest and most ancient visible stratified rock is the jurassic limestone, which constitutes the central mass of the axial, and all the higher lateral chains.

Lithologically, it is usually in heavy and well-marked beds, the line of strike being very commonly in the general direction of the chain, and the beds tilted to a high angle, so that a very large proportion of the whole mountainous surface of the country, consists of highly inclined beds, running about north and south. There are, however, large exceptions to this: in the mountain knot, of Muro and Bella, amongst other places, for example, the beds, nearly vertical, often cross the lines of valley at right angles. Again, in the great range of La Scorza, or Monte Albano,—south of the Salaris, and between the Val di Diano and the Plain of Pæstum—the beds support a large elevated and nearly level table land, with an east and west strike, and inclined at various angles dipping to the south, and are piled up fully 3,000 feet above the valley of the Rio Negro. They seem to dip inwards, towards the centre of the table on top, so as to rampart it all round: it is the largest surface of mountain table land in the kingdom.

Everywhere this lower limestone presents traces, of immense disturbance and dislocation, and of enormous denudation.

Its colour is most commonly yellowish ash gray, and

when most compact, it has quite a liassic aspect in hand specimens; it varies much in colour, however; red, purple, variegated, and nearly white, are to be found. In many places it presents metamorphic characters, and becomes for limited areas, flinty and hard.

A great band of this limestone, extends from around the Terra di Lavoro, southwards to the Gulf of Taranto, from thirty to forty miles wide on the west side; it winds about, forming the summits of all the hills that rise out of the level bed of tufa, that surrounds Naples, and then stretches away northward in a still wider band to the eastward, and a narrower one to the westward, sides of the peninsula. A large region from Barletta to Gioia, on the Adriatic coast, also consists of it. Within the first-mentioned band, and resting upon this limestone, are scattered immense deposits of a coarse calcareous breccia, consisting of rounded masses of various sizes, (sometimes, as north of Potenza, very large, reaching eight or ten feet in diameter,) and cemented together, with similar but softer material, in ill-defined heavy beds, usually much less inclined, than those of the limestone beneath. Whether this rock belongs to the cretaceous series, or to what other, I am unable to say: it occupies the bottom of many of the narrow valleys, and in one place on Collegno's map, appears to be assigned to the pliocene tertiaries, but probably in error.

Above the lower limestone, reposing upon it, laid against its highly-inclined beds, and often mixed with it, in perplexing confusion—we find the nummulitic and hippurite limestones of the cretaceous formation, always characterized to the eye, even far away, by the want of clear bedding, the more rounded outlines, of the lower mountains composed of it, and by its brighter colour. It is usually nearly white, often sufficiently hard and dense to work well, as a beautiful building material, capable of a good polish; but also passing insensibly, within a few miles, into a soft, sandy stuff, of little coherence, like a compound of English chalk, and fine white Dorsetshire sand, but still forming rocky eminences several hundred feet in height.

Upon this again appear to lie, chiefly in the bottoms and on the flanks of the valleys, beds of marls of various tints, of enormous thickness—600 or 700 feet in some places. These I presume to be the sub-Apennine marls of Collegno.

In the lowermost portion of these marls, beds of yellow and brown sandstone occur, here and there of great thickness; in some places, they are traversed by beds of indurated, highly ferruginous and magnetic, dark gray calcareous rock,—by beds of gypsum,—and in very many places, give evidence of metamorphism. The beds, usually soft and sectile, and acted on with immense rapidity by river erosion, being converted into masses of striped jasper, often of great beauty and extreme hardness.

In the neighbourhood of Potenza, there are large developments of indurated argillaceous slaty beds, of dark blue gray colour, which Collegno appears to refer to these sub-Apennine marls, but which appeared to me widely different, in lithological character at least, from those last referred to.

Above these marls, reposes, the usually great depth of alluvial clays, which constitute the valley bottoms.

The tops and flanks, of the upper and lower limestone mountains, are almost always nearly bare of soil; it has been all swept off from them and levelled under water in the valley bottoms, to form these rich plains of agriculture. The soil is commonly a dense tenacious brown or red brown loam, often free from stone for large areas, and composed of the detritus of the calcareous and argillaceous rocks. That combining the sand of the cretaceous rocks, and the clay of the marl beds, is of high fertility.

Where the fall is small, the rivers (as the Tanagro in the Val di Diano) run slowly, upon beds of these clays, and form for themselves a permanent bottom, paved by the angular fragments and boulders of limestone washed from the soil; but when the fall is rapid, as in the Rio Agri—a view in the valley of which, is given in Photog. No. 110 (Coll. Roy. Soc.), then erosion takes place upon a scale of great grandeur. The alluvial soil is cut through to the bottom; the marl beds follow it; and the river runs upon a bed of loose blocks of limestone, resting on the bare rock, at the bottom of a ravine or "nullah," with steep sloping banks, which are continually washing away and slipping in, and at a level often of 300 to 700 feet below the flat of the piano, or valley bottom, that spreads out above, for miles at either side. The rocks in the foreground of this photograph, are of the calcareous breccia described. The rivers themselves, fed by the heavy rains of the wet season, which fall with something of the regularity and violence of the tropical rainy seasons, by the wet of winter, and by the sudden melting off, of the vast masses of snow that fall on the higher ridges, (and at elevations exceeding 2,000 feet above the sea, lie accumulating during winter,) are of a torrential character, and almost all flow thick, and turbid with brown alluvial sediment, to the sea.

North of the great transverse ridge about Atella, we begin to encounter the great tufa deposit, of the ancient volcanic region around Monte Vulture, and extending northward to beyond Melfi, where it is evident that denudation, has levelled and spread out, over the limestone, prodigious quantities of the tufas and decomposed lavas, at a period probably long anterior to the ejection, of the earliest of the tufas of the Vesuvian tract, and where torrent and rain erosion, present features of the largest, and most instructive character.

Details, either of that, or of the Vesuvian volcanic tract, are beside our present purpose, however, except to remark, that in both, the limestone laps in under the superficial volcanic products to an immense distance; indeed, in the Terra di Lavoro—as we travel, for example, from Naples to Caserta and Capua—it is obvious that the level plain of tufa over which we pass, out of which the limestone mountains rise sheer and abrupt on all sides, has been run in and levelled between them, and has traced out by its surface, the contour, of every sinuosity of their narrow winding valleys, when all were under water, and that the limestone of the hills, in reality underlies almost the whole of the great tufa bed. Earthquake vibrations, therefore, penetrate both these volcanic regions, through the intervention of the harder and more elastic limestone beneath, the tufas being thus shaken like plastic clay in a saucer, just as the great alluvial beds in the more southern valleys, are shaken by the vibrations, primarily propagated through the limestone surrounding them. Of the two, probably the tufa is the worse material, for the easy propagation of earth wave.

The connection between one valley and another, at a not greatly different level, is not unfrequently through a gorge or serrated cleft, fractured through the rock, in the bottom of which a torrent roars, while the road or mule tract is over the shoulder above; such are Campostrina, between the valleys of the Calore and Tanagro—and that of the Gioija, at Muro. In general, however, the valley bottoms are at different levels, and are reached by passing over low shoulders between them, the streams finding vent in sinuous, but not very deep ravines.

The lower mountain ranges, of the cretaceous limestone are not very steep, though the slopes can nowhere, except in the rolling plain country, be called gentle. The higher ridges, are always steep, and frequently characterized by a shaggy bristling grandeur, of crest and outline, greater than one is prepared to find in mountains, of a formation so recent.

The fall of the rivers generally, throughout the kingdom is rapid, the mean breadth of the land, not giving an average length of bed, of much above a hundred miles, in which the average fall is probably above 3,000 feet; and even the great rivers, have their volume so augmented in winter, that on reaching the seaboard plains, their velocity is still very great, on debouching into the sea. Thus, the Salaris, after having traversed the plain of Pæstum, retains a mid-surface, winter velocity, of about eight feet per second.

The towns in the earthquake region to which this Report refers, are nearly all built, as stated, upon rocky eminences, within the mountainous region; in some cases, however, within it, they are (or were) built upon the alluvial clay deposit, on the level of their respective Piani.

In the Capitanatas, Basilicatas, and Bari, there are many towns—some of them large and important, that stand upon the plain, or on elevated knolls upon it—none, with the exception of some on the coasts, such as Amalfi, are found nestling into valley bottoms of small size, as in colder climates, and such towns are usually of extreme antiquity.

In the Appendix (No. 1) to this part, a translation is given, of that portion of the report of Professors Palmieri and Scacchi, on the Melfi earthquake of 1851, which comprises their general account of the geology of Southern Italy. Although still leaving much to be desired, it is the best sketch I have been able to meet with.

A few monographic memoirs on Neapolitan geology exist, such as Elie de Beaumont's, and others, on the Lignite formation of Calabria ('Comptes Rendus,' 1858).

Collegno's 'Elemente di Geologia Practica e Theoretica,' Torino, 1847, contains a good deal of information as to the geology of Central and Northern Italy, with sections of the Northern Apennines; and so also does Pilla's 'Saggio Comparativo die Terreni d'Italia,' and his ' Trattato di Geologia' (Pisa, 1847–51); but for Southern Italy I have met with no corresponding information.

APPENDIX TO PART I.

No. (I.)

Translation of the general account of the Geology of Southern Italy, forming Chapter I. Part I. of Profs. Palmieri and Scacchi's account of the Earthquake of 14th August, 1851.

'*Della Regioni Vulcanica del Monte Vulturi, e del Tremuoto ivi Avvenuto nel de 14 Agosto, 1851, relazione fatta per incarico della Reale Accademia delle Scienze, da Liugi Palmieri ed Archangelo Scacchi.*' Napoli. Gaetano Nobili, 1852.

OF THE NEPTUNIAN ROCKS THROUGH WHICH THE VOLCANOES OF VULTURE FORCED THEIR WAY.

THE ancient fires of the region of Vulture opened a road for themselves through the neptunian rocks, which are not materially different from those which surround the other volcanic districts of Campania. The examination which we have made of the neptunian rocks of our kingdom from Balzorano, in the southern limits of Abruzzi, to Tarentum in Puglia, and Pizzo in Calabria, has proved to us that the same rocks occur everywhere in the same order, and with nearly the same topographical conditions. Geologists who have visited these countries have met with no slight difficulty in establishing a division in these sedimentary rocks, according to the order in which they have been deposited; and the difficulty has not been diminished by comparing them with rocks of determined epoch, which might be contemporary with ours. For us, who have not had the advantage of seeing in their natural position any of the various sedimentary formations (except those of the kingdom of Naples), the difficulty has been still greater, nor can we flatter ourselves that we have surmounted it. Meanwhile, without entering into discussions which would be foreign to the principal aim of our work, but keeping to what appears to us to agree with our observations, we prefer to divide our neptunian rocks into three series, that is, three distinct formations. In the first series we shall include all those calcareous rocks which are particularly characterized by Nummulites, Nerineæ, and those organic forms of which we have no example in the fauna of the present epoch, and which paleontologists, uncertain of their real nature, have denominated rudimentary. (*Rudisti.*) The greater part of our Apennine mountains being formed of this calcareous rock, we shall retain for it exclusively the name of (*Calcarea Apennina*) Apennine limestone. The rocks of the second series, very varied in their mineralogical composition, agree in being distinctly stratified, in being almost entirely destitute of animal fossils, and in sometimes containing a great quantity of vegetable fossils of the Fucoid order. Although it is not easy to find Fuciform impressions everywhere in rocks of this formation, nothing better indicates their character than the presence of these plants; for the absence or extreme rarity of animal fossils is a negative characteristic which in exceptional cases may be affirmed of every species of rock. The last series comprehends marls, limestone, and sandstone, abounding in marine fossils, the greater number of which belong to species which now exist in our seas. To these rocks we shall confine the name sub-Apennine, although others elsewhere have designated some of the rocks of the preceding series with this denomination. They undoubtedly belong to the supercretaceous period; and if the igneous phenomena of Epomeo in the Phlegrean region, as we have shown elsewhere,* have pre-

* Scacchi. 'Geological Memoir of Campania.' Naples: 1849. Pp. 19, 20.

ceded some nearly supercretaceous deposits; yet in the Vulture district, where the order of position between the neptunian and volcanic rocks is in several places very easily seen, I have never found the igneous rocks, lying beneath any of the various rocks of the third series; so that we may maintain that the emergence from the sea of the sub-Apennine deposits, took place *previously* to the first fires of the Vulture. There are also in this same volcanic region frequent and extensive fresh-water formations; but, contrary to what has been observed of the marine deposits, they lie invariably *above* the rocks of igneous origin, and closely resemble the sedimentary deposits which are in process of deposition in our own days, under very restricted conditions, in like places abounding in water. Pursuing a chronological order, we shall briefly speak of these when we have concluded our remarks on the volcanoes of the Vulture district.

First Series.—APENNINE LIMESTONE.

The rocks of which the oldest formations of aqueous origin in the kingdom of Naples are composed, are almost exclusively calcareous, and include many varieties, which seldom constitute an essential difference. The most frequent of these varieties is the compact, with conchoidal fracture, of a white, or clear smoke-gray colour. Another variety, somewhat less abundant, has a granular texture more or less distinct, in which oftener than in the preceding, small cavities lined with crystals of the same substance are found. Passing over several varieties of minor importance, we shall enumerate four others. The first has a brecciated structure, enlivened by gay colours, and is capable of receiving a beautiful polish. Of this we have magnificent specimens in the marbles of Vitulano, and Mondragone, in Terra di Lavoro. The second is of a beautiful white colour, and pulverizes easily at a touch. It may be seen between Piedimonte di Alife and S. Potito at the foot of Matese in the district of Melfi, after the 77th milestone on the road from Valva, and in many other places. The third variety, not very different from the preceding in appearance, is that which geologists denominate chalk, and which, so far as we know, is only found at Monte

Gargano*. The fourth is bituminous, and is met with in many places, particularly in those which abound with Ichthyolitic fossils. Quartz or Firestone (*Piromaco*) is so frequently found in limestone of this series, that it may be considered characteristic of it. Sometimes it appears in veins at the junction of the thicker calcareous strata, but more frequently it is imbedded in them, assuming various figures, among which the spherical is remarkably perfect. Astonishment at the perfection of the globular figure of these *Rognoni di Piromaco* † has occasionally given rise to the foolish supposition that they are areolites. The manner in which this variety of quartz is found in the limestone clearly proves it to have had the same origin as the rock which contains it; and this is corroborated by the frequent cases in which it partly or entirely takes the place of the carbonate of lime, of which the fossils enclosed in the same rock were originally formed. In such cases it is worthy of note, that the other variety of crystallized quartz is occasionally united to the Piromaco. With our present chemical knowledge, we can easily understand that silex, as well as carbonate of lime, may have been held in solution in the water by which they were deposited. But we cannot, with the same facility, account for the silex being precipitated in the state of Piromaco, or as crystallized quartz, whilst we know that silex is naturally deposited by some mineral springs, or artificially by solution in concentrated waters, with the qualities proper to *Gieserite*, or hydrate of silex. Perhaps arguing from the frequency with which fossils are found converted into Piromaco, we may attribute its presence to organic substances, as Becquerel observed that organic substances in a state of putrefaction, deposit crystals of pyrites in a solution of sulphate of iron.

Carbonate of magnesia, in widely-varying proportions, is always united to carbonated Apennine limestone. The large quantity which is occasionally found, we think, must be attributed to the facility with which, in some places more than others, this rock undergoes great changes, caused by the continual influence of

---

* Limestone passing into chalk is found abundantly in many other places, especially in the lateral valleys to the westward of Padula.— R. M.

† Kidney-formed masses.

the weather (meteori). One of the most surprising examples of this kind of decay may be observed in the upper part of the valley of Tramonti, in the province of Salerno, where the phenomenon extends over a large tract of ground. In some places, as in the vicinity of Amalfi, the magnesian limestone presents the characteristics of Dolomite, having a granular texture, and dissolving slowly in acids. However, we cannot subscribe to the opinion of those who attribute the presence of magnesia to the action of internal plutonic masses, which have given to the sedimentary limestone rocks the characteristics of Dolomite. There is no ground for refutation of the opinion, that true Dolomite is of neptunian origin, and therefore its presence is insufficient to prove that the change has been occasioned by plutonic rocks. And the reasons which prevent our applying the theory of dolomization to our Apennine mountains are—1st. The absence of plutonic rocks in their vicinity, which might occasion the phenomenon. 2nd. The rare occurrence of true Dolomite, and the fact that it is confusedly mixed with magnesian limestone, which does not possess the distinctive characteristics of Dolomite. And, lastly, the organic forms of well-preserved fossils in those rocks which most manifest dolomitic qualities. The last is one of the best arguments against the maintenance of the Dolomite theory in reference to our magnesian limestones, but of it we do not find many examples. We might conclude that the presence in great quantity of carbonate of magnesia is adverse to the propagation of marine animals, or that it has contributed to the destruction of the shells of the ancient fauna. In the Dolomite of Amalfi, where we found several well-preserved casts of Terrebratulæ and Ammonites, this opinion is still further strengthened; for in all the examples which we examined their shells were entirely destroyed. The preceding considerations, however adverse to the phenomenon of dolomization in the rocks of our Apennines, do not exclude the internal plutonic forces, by which they were disturbed from their primitive position, and raised to their present height; nor the phenomenon of metamorphosis, which is clearly manifested in some particular districts, where limestone is found at a short distance from crystalline rocks of igneous origin. In the district of Castrovillari, perhaps more than anywhere else, frequent examples of metamorphosed Apennine limestone are met with. It acquires a granular crystalline texture, not unlike statuary marble, and contains passing through it some thin veins of quartz, with chlorite, many crystals of pyrites, and sometimes cinnabar, and the traces of fossils are very nearly obliterated. There are yet more evident proofs of metamorphosis in the rocks of the succeeding series, which are changed into Quartzite, Stannite, Talc, and schistose argil, all of which are found in the limestone of the region. The Apennine limestone has generally been reported poor in organic remains, which might furnish paleontologists with determined characteristics. Our researches lead us to a contrary opinion; but as we cannot, without deviating from the design of our work, enter into an examination of the different species of fossils which we have found, it must suffice to say, that the number of species, and the quantity of each kind, are such as to give us a correct idea of the abundant fauna of the sea, in whose depths our highest mountains were formed. Any who wish for a convincing proof of this assertion may find it in the rich collections which, within the last few years, have been deposited in the Mineralogical Museum of the Royal University of Naples. The most frequent and abundant organic forms which we have found belong to the *Rudisti*, and of these we may say, that there is no place in which we cannot discern their trace; and in some spots, whether owing to the nature of the rock in which the fossils are preserved, or from the habits of those animals to live in myriads in a small space, the quantity is so enormous, that the entire rock appears to be composed of them. Monte Gargano, Monte Lesule, in Matese, the brown limestone of Il Ponto Consolazione, near Lauria, in Basilicata, present striking examples. Although they are usually so completely petrified, and identified with the rock itself, so as to render the determination of the species difficult, yet by attentive examination we can distinguish a difference between those found in places not very distant from each other. The large species of Nummulites next claim our attention; and while, like the *Rudisti*, or even more strikingly, they are found united together in myriads in one place, they are not, like them, distributed over a large district. In many extensive districts they are in vain sought for. Besides, in Monte Gargano, and the neighbouring islands of Tremiti, renowned for fossils of this kind, no contemptible deposits are found in the vicinity of Lama, in Abruzzi, near Casalbone, and not far from Ariano, in the province of Avellino, and in the territory of Benevento (Olivella di Pacca), where they are disseminated through limestone and Piromaco. The nerinæa, another kind of mollusc, which deserves our attention, usually accompanies the *Rudisti*, and is found in many places; several species, differing in size and form, being recognizable. Besides these three groups of fossils—to mention but a few examples of a long series, containing many species, some of which probably ought to constitute new genera—we must record, in the province of Terra di Lavoro, the spiral shells of the *Lumachelle* of Monte Casino and Vitulano, the *Diceratiti* of Monte Licinio, near Cerreto, and some shells, allied to the *Natiche*, of Monte Lesule, in Matese. In Monte Gargano there are also some remarkable impressions of plants, probably of the family of the conifers, two gigantic specimens of *Bulle*, a Pirula, and Ammonites Rothemagensis; and in this same mountain, as well as in the vicinity of Amalfi, and near Castelgrande, in the district of Melfi, different species of zoophytes are not scarce. As to our ichthyolites, of which but few species were known before the recent works of Costa (Costa, 'Paleontology of the Kingdom of Naples, 1850'), we are now acquainted with a great many in the mountains of Pietraroia, Giffuni, and Castellamare; and by the discoveries also of Professor Costa, of which as yet only a brief account has been published (Costa, 'Hints relative to the Discoveries of the Paleontology of this Kingdom made in the year 1851'), we have made out at Pietraroia, in the icthyolitic limestone, some species of reptiles. From the brief account we have given of the fossils found in the Apennine limestone, it is easy enough to infer that all, or at least the greater part of them, belong to the great Chalk formation. But from such observations as have hitherto been made, although we cannot say that they are sufficiently numerous, it is not equally easy to establish any division among them; nor can we decide with any certainty whether some of them may not belong to still more ancient formations. Yet, not knowing of any instance in which *Rudisti* have been found in those rocks which contain the fossil fish at Giffuni, Pietraroia, and Castellamare, we are inclined to think that they ought to be placed in the Jurassic group. Meanwhile we consider it better to refrain from putting forward ill-grounded opinions, and to defer the solution of this question until a future day shall bring more decided facts to light; for at present, so far as our knowledge extends, ichthyolitic limestone has not undoubtedly been found either above or below that which contains *Rudisti*, nor have any of our species of fossil fish been recognized in rocks of a well-determined epoch, although some species have been found in the chalk formation of Monte Gargano. The topographical character of mountains formed of Apennine limestone enables us to distinguish them at a great distance: their long narrow summits, the lesser ramifications branching off from their sides, and dipping down till they end in an acute ridge, their slope not unfrequently broken by majestic steps, against which the inferior rocks lean, as if against a firm vertical wall, the strata of which they are composed appearing gradually to tilt up, as if they would touch with their ends the high acclivities, and even the most elevated ridges, form so characteristic an aspect, that they are easily distinguished from neighbouring mountains and hills of a different nature. Admitting a few unimportant exceptions to this general order of topographical configuration, the fact of greatest significance to which it is necessary to direct attention is, that the same Apennine limestone which is visibly displayed to a great extent in the provinces of Capitanata, Bari, and Lecce, there assumes a completely different aspect. We find low hills, commonly called *Murge*, in the province of Bari, extending in various directions, and sloping down to a vast plain, which with the mountainous region forms a winding line from north-west to south-east, almost parallel with the coast of the Adriatic, that extends from the Gulf of Manfredonia to Brindisi. To the characteristic outline essential to the *Murge* we must add the no less important character apparent in the arrangement of the strata, which, besides being more distinct than is usually observed in mountains, is generally horizontal, or somewhat inclined to the horizon. These differences between the perpendicular strata of the mountainous regions, and the horizontal strata of the low hills, sufficiently prove the elevation, or at least the great displacement, to which the former were subjected; while the latter have preserved the same position in which they were originally deposited, relatively at least to the horizon, if not to their distance from the centre of the earth. It might be thought that the limestone of the *Murge*, the stratification of which is so different from that of the mountains, was deposited at a later period; and this would appear to verify the opinions of those geologists who, unacquainted with the fossils it contains, considered it to belong to the super-cretaceous period. But the frequent occurrence of Hippurites in it contradicts this opinion, and as we find it agrees in palæontological characteristics with the limestone of the Apennines, we are compelled to hold them contemporary. Nor can we find any other reason for the difference of stratification except that now mentioned, viz., that the first was not subjected to the internal plutonic forces of our planet.

Second Series.—Rocks with Fucoids, or the Macigno Formation.

Generally speaking, their topographical aspect sufficiently distinguishes the rocks of this series from those of the preceding; and, although they are often found in the elevated regions of the Apennines, yet they never form great chains of mountains. Their usual appearance is that of small mountains or hills, with rounded and depressed summits, and in a few cases, where the strata are unusually thick, and greatly elevated, they assume the appearance of limestone mountains. The old summit on which the city of Monteverde is built, and which rises much above the lesser prominences which surround it, furnishes us with a most striking example of this effect; and even here, though the developed form of the mountain is an exception to the general rule, yet its height does not equal that of the ordinary limestone mountain masses. The different species of rocks which compose this formation are remarkable for the manner in which they are stratified, frequently alternating with each other, and presenting the most beautiful appearance of regularly disposed strata, the one kind surprisingly distinct from the other. The ordinary thickness of the strata varies from a decimeter to half a meter. Instances of great thickness are less frequent, and occasionally the layers are very thin. The most important point observable in these rocks is their strange position, so highly inclined to the horizon. The distinct development of the strata enables us to determine their degree of inclination and direction with great accuracy; yet as to their direction we have not been able to ascertain towards what point they are generally elevated, the direction of their inclination being found very various even in places separated by short distances. The angle of inclination also is very uncertain, varying often between 25 and 50 degrees; nor are instances rare in which it attains to 70 or more degrees, and in some places the strata might be called vertical. In no other of our sedimentary rocks have we better evidence of the great displacement which must have occurred, than in that which the great inclination of these strata presents, as it is impossible to conceive how they could have been deposited as they now appear. Yet it is in rocks of this series that we find the greatest difficulty in assigning the true cause of this fact. First of all, finding several strata of clay which crumble easily, on account of the softness which they acquire from the absorption of water, and the limestone and sandstone strata having, on account of their thinness, but little resistance, we are justly inclined to think that the elevation of the strata is the effect of the rents occasioned by the continual mining of subterranean water. Admitting the possibility of this reason, and maintaining that it may have in some measure effected the change of the primitive situation of the stratified rocks of this series, it is not probable that it alone, could have produced such great effects, and often in an uniform manner over a great space, as observation has made known to us. In many places it is entirely impossible that any other cause than that of plutonic forces, having their seat at a great depth beneath the terrestrial surface, could have, so strangely and throughout entire regions, elevated the enormous stratified masses, and thus discovered their internal structure; but for which we should never have been conveniently enabled to examine them, and perhaps might still have remained in ignorance. And now, for a clear explanation how catastrophes of a like nature appear to us to have happened, it is necessary to declare, that if, by relating facts as they first appear to sense, we have attributed the actual position of strata inclined to the horizon to elevation, this does not exclude the idea of undermining operations. And since, whether they have been elevated or whether they have been undermined, we should always find their primitive horizontal position deranged in the same manner, it is not easy to decide which of the two conditions has rendered the region mountainous, or, if both were in operation, which of the two has had the greater influence. For if we consider the general physical property of matter to diminish in volume by the decrease of its temperature, the theory at present held by all, of the primitive state of igneous fusion of our planet, leads us to the necessary consequence, that in cooling it must have diminished in size, and the forces which have acted on its consolidated crust must induce it to approach the centre. In the district occupied by the volcanoes of the Vulture, other particular considerations present themselves to the mind of the geologist who contemplates the strata of the neptunian rocks, here, perhaps, more elevated than elsewhere. As we shall presently state, these volcanoes are surrounded on every side by hills, formed of rocks of the second and third series, which must have felt the effects of the disturbing volcanic eruptions, and therefore it is easy to attribute their elevation to the same eruptions. In the east side of the base of Mount Vulture, along the line occupied by the cities of Melfi, Rapolla, Barili, and Rionero, in many places it is easy to observe the strata of limestone and red marl beneath the lavas and volcanic conglomerates, with the ordinary character of elevation proper to this formation. Half way on the road between Rionero and Barili, on the left-hand side going from Rionero, in the place denominated the *Valle del Salice*, a long series of the outcrops of stratified rocks, which form the bed of a brook which flows over them, is seen. The average of their inclination, though variable, is about 70 degrees, and they are elevated from the side turned towards the east and south-east. It is enough to say that they are inclined in a contrary direction to the slope of the Vulture, to prove it impossible to attribute their elevation to an internal force, whose centre of action should coincide with the central part of the volcano. The same thing may be observed in many places on the same road, so that this arrangement of strata is invariable for upwards of a mile. And to this fact surely Fonseca alludes, when he says that between Rionero and Barili the limestone strata are inclined almost at right angles to the declivity of the mountain. (Fonseca, 'Geognostic Observations on the Vulture.') Along a brook which runs near Barili, from the north-west side, the same strata of limestone and marl on the right bank are elevated towards the west 44 degrees, and on the left bank, towards the north, 31 degrees; and, one being almost opposite the other, their displacement cannot be referred to the action of the Vulture. In a valley situated to the north of Rapolla, the waters which run through it pass over limestone strata; and in one place where these are very apparent we have found them elevated 55 degrees towards the north-east, not to mention some which are even nearer to the vertical position. Lastly, omitting other facts of a similar nature which are less conspicuous, along the little river which runs round the hill of Melfi in the north-west side, some strata of grey argil are seen, which have the same inclination as the hill; and in the west side, not very distant from the bridge, commonly called Gaetaniello, there are some strata of red marl, containing fucoids, inclined 70 degrees towards the north-east. Their elevation cannot be attributed to the principal volcanoes of the Vulture; nor does it agree with the centre of action of the volcano of Melfi, as then they would have been inclined to the east. From these facts it is natural to conclude that the displacements observable in the neptunian rocks of the volcanic region of Vulture were not occasioned by the same forces which gave rise to the volcanoes, but rather that they occurred previously to their eruptions. This opinion is strengthened by the fact, that they present the same appearances as have been observed in other places, where the distance from the Vulture and the particular manner of dislocation prove that the eruptive volcanic force could not have occasioned such displacements. We shall select, from many instances which we might quote, that of the lofty eminence which we recently mentioned, on which the city of Monteverde is built. It lies to the north-west, distant in a direct line little more than four miles from the lakes of Monticchio. It is chiefly composed of large strata of Macigno, some of which are upwards of

four metres in diameter; and in the southern summit, called *Sierra della Croce*, the strata being laid bare by several hollows, show an inclination of 47 degrees; and from this side rises another opposite summit, having the same inclination, from which we may conjecture that the Sierra della Croce has been separated from it. Therefore, even supposing we admit that the internal impulse accompanying the fires of the Vulture, having its centre in the neighbouring lakes of Monticchio, may have extended to a great distance, it is impossible that the actual arrangement of stratification of the Macigno of Monteverde could have been occasioned by it: an order not essentially different from that of the stratified rocks which are displayed from Melfi to Rionero.

In the district of the Vulture, much better than in any other volcanic region of our kingdom, we can observe the manner in which the volcanic rocks interstratify with the neptunian; and, comparing them with the observations here collected, we cannot reconcile them with the idea that volcanic forces could have had so extensive a field of action near the terrestrial surface. On the contrary, we are led to the opposite opinion, namely, that the space is very limited in which volcanic explosions can occasion elevations or other perturbations of ground, and that the first convulsions are almost always concocted under the materials which are subsequently ejected. As we must return to this argument in another part of our work, what we have already said is sufficient to testify, that in our opinion the volcanoes of the Vulture have had no part in the elevation of the rocks of the second series, in which they have appeared.

The relation of arrangement between these rocks and the Apennine limestone presents another field of inquiry, in which it is not easy to see clearly. Is it beyond a doubt that the former belong to a period subsequent to the latter? Are the elevations of the first cotemporary with the elevation of the second? Have they been once or oftener convulsed? What difference of conditions results from the difference of composition between the rocks of the first and second series? These are the principal questions which the geologist is compelled to discuss. We shall speak of the last when we have given the necessary mineralogical description of the numerous fucoidal rocks. As to the first question, we must admit that, throughout Campania and the greater part of the Principatus, we cannot, generally speaking, perceive with sufficient clearness the super-position of one system of rocks above the other. Not so in Lucania: there the super-position of the fucoid strata above the Apennine limestone is very evident, and their conditions are notably different. These observations have induced us to maintain, not only that the former were laid down at a subsequent period to the latter, but also, what is still more important, that they belong to two distinct formations. Perhaps the most suitable place for examining these conditions is the Valva Road, along which, from Oliveto to within a few miles of Atella, we never lose sight of the line of contact between the hills of the second series and the mountains of the first. The latter appear to come out from under the hills which lie round their bases; and in some places, as at Fontana della Rosa, between Laviano and Muro, the order of the strata which form the hills is clearly seen to rest upon the Apennine limestone. At the same time we can observe the difference of direction and inclination of the strata belonging to the two systems: a discordance which is also manifested in the different topographical aspect of which we have already spoken, and which cannot exclusively depend on difference in mineralogical composition of the rocks. As the strata of clays, sandstones, and limestones with fucoids, in the northern provinces, do not possess any notable difference of composition from those in the southern, nor are the paleontological characteristics at all different, arguing from analogy, we consider them all to belong to the same formation, and are confirmed in this opinion by never having met with any fact which could clearly contradict it. From this difference between the Apennine limestone and the rocks of the second series we are led to infer, as a necessary consequence, that the former must have been displaced before the latter were deposited; and again, these latter rocks being so much inclined is the proof of a second period of elevation. The manner in which the rocks of both series inter-stratify with each other appears to us sufficiently to declare that the more ancient formations cannot of necessity have been exempt from the disturbing force which displaced the more recent. Of these general conjectures relative to the convulsions to which our sedimentary rocks have been subjected we find satisfactory proofs in the southern regions; whilst, on the contrary, in the northern, as we have already observed, the relation of arrangement between the rocks of both systems is not manifested with sufficient clearness, nor can we support the theory that the same phenomena have everywhere occurred. Having already observed that the Apennine limestone may be divided into two distinct regions, the one mountainous, the other almost level, called *Murge*, a fact of no slight importance appeals to our consideration, relative to the distribution of rocks of the second series in connection with this division, viz., that they only occur in the mountainous region. It was not without surprise that we traversed the province of Bari, and the neighbouring districts of the provinces of Lecce and Capitanata, seeking diligently, without success, for rocks of the second series, of which we never found a trace. We must confess that we cannot clearly account for this circumstance. Where the mountains formed of Apennine limestone rise, there must certainly have been during the epoch in which the rocks of the second series were deposited, a topographical condition completely different from the level plains. Ought this topographical difference to be considered sufficient to prescribe the limits within which such rocks can be formed? Now, as we shall presently show, among the components of these rocks, there are some which have very probably been transported from the granite mountains of Calabria, and in general consist of materials which may have come from distant places. Ought we to regard the direction in which their elements were transported as the cause of their being found in certain regions, and being wanting in others? The mineralogical composition of the fucoidal rocks is extremely varied. We may divide them into five different species, viz., limestone, marl, sandstone, limonite, and gypsum; and each of these admits of being subdivided into many varieties, of which we shall only record the principal. The varieties of most frequent occurrence in the limestone are the marly, of various colours, sometimes with those beautiful appearances which take the designs of ruined buildings (*calcarea ruiniforme*). Another, less frequent, but not less characteristic, is a breccia, of very minute fragments, with rose-coloured cement, sometimes intense, and again diffused, bearing a great resemblance to red porphyry, consequently of lovely effect in workmanship. The most beautiful specimens of the *calcarea ruiniforme* are to be found in the neighbourhood of Gesualdo and Frigento, in the province of Avellino, and the second variety is more common in the district of Melfi than anywhere else. If the limestone of this series contains almost invariably some clay, the clays, on the other hand, are always mixed with some proportion of carbonate of lime, which gives them a marly character. They are sometimes compact (amorphous), but more frequently divided into thin laminæ, without losing the property of forming with water a ductile paste. They are usually of a gray sky colour, and in Lucania are not unfrequently red. These rocks on one side pass by imperceptible degrees into limestone; on the other beginning to contain minute particles of mica and small grains of sand, the latter gradually becoming more abundant, they change into sandstone. The sandstone itself, owing to the size of the grains of quartz, and their abundance or scarcity, and also to their varying degree of tenacity, present innumerable differences, which are of no importance. They possess, for the most part, the characteristics of true Macigno; in some cases are good for sharpening edged tools, and in others may be used with advantage for making bricks or crucibles capable of bearing a high temperature. The Limonite is seldom found pure, and its deposits are so scarce, that they can with difficulty be profitably used for the extraction of iron; nevertheless, mixed with carbonate of lime, it is rather frequently found in each of the three preceding kinds of rocks, especially in the limestones and marls.

Ferruginous sandstones, occasionally mixed with deposits of Limonite, are not scarce in the district of the Vulture, and it is requisite to take care not to confuse them with volcanic productions. When the limonite unites with marl, it occasions such strange forms, that the naturalist no less than the uninitiated must regard them with astonishment. Besides what I have said relative to these *Eagle stones*, of which beautiful specimens are to be found in the district of Gerace, there are some remarkable varieties in the Fucino district near Pietraroia, which, from their resemblance to pieces of petrified serpents, have been denominated *serpentini*. To these kinds of configuration we may add the spheroidical masses with laminated concentric structure, found in the vicinity of Alberona, in Capitanata, and the prismatic forms of the valley of Ansanto, outwardly composed of large crusts of Limonite with *Siderosa*, and filled interiorly with marl, and often with pieces of the same Limonite. In the sandstone, besides the minute grains and rounded pebbles (*pezzetti rotolati*) of quartz, of which it is essentially composed, pebbles of rock crystal are sometimes found in great abundance. For the most part they belong to the granite, quartzite, or porphyry, and vary greatly in size, increasing from that of a filbert to about two decimeters in diameter. In occasional instances they are found of a surprising magnitude; but these, as well as the smaller specimens, have a rounded surface. We may instance one, which was found above Monte Vergine, near Avellino, more than five decimeters in diameter; and another, in the region called Fontana delle Rose, not far from Muro, whose greatest diameter was sixty-three centimeters. This last region, made known by the published works of Tenore and Gussone,* ought to be visited in preference to any other, by those who desire to examine the great rolled masses of granite which exist in our Apennines. On the road from Laviano to Atella, a little past the seventy-first milestone, we met on the right-hand side, a path leading to a spacious valley, through which the waters of the Fontana delle Rose run. Along this rough path we often meet with large granite boulders, and in the valley following the course of the stream, several large isolated ones may be found. Of these last there can be no doubt, that, like those still imbedded in the sandstone, they too were inclosed in the rock, and when it was disintegrated they remained scattered as we see them through the valley. Many such rolled masses of granite, and of another kind of crystalline rock, are found along the River Olivento, commencing at the source under Ripa Candida, and extending to its junction with the torrent of Macera; similar rocks are also found in other places surrounding the volcanic region of Vulture, or on the Vulture itself. We have seen

* Tenore and Gussone. 'Memoirs of Tours performed in the Years 1834-1838.' Naples: 1842. Pp. 75, 76.

some at the Bosco di Gaudianella. When we have explained the relation of arrangements between the neptunian rocks of the third series and the volcanic, it will not be difficult to understand how they came there. Lastly, among the districts in which numerous pebbles of rock crystal are observed, we must enumerate the vicinity of Pietraroia, remarkable for the great variety which is found there. Before we had observed the granites enclosed in the limestone of the Fontana delle Rose, we referred the blocks of the same kind found at Monte Vergine, and Pietraroia to the doubtful series of masses of similar rocks, sometimes of enormous size, called by geologists (*erratici massi*) erratic blocks.* We now class them in another group of rocks, and assert that their origin is not at all different from that of the minute grains of quartz, of which the Macigno of the fucoidal rocks is composed ; the grains as well as the spangles of mica which are frequently seen in clay, being minute particles of granite, or of some other crystalline rock. The greater number if not all the varieties of granite found in isolated blocks among the sedimentary rocks of the Apennines, resemble, in the most minute particulars of their sensible properties, the rocks of the same kind which we observed in their primitive arrangement in Calabria. If this is sufficient to assure to us that they owe their origin to the granitic mountains of Calabria, it will follow, that we must hold, that at least the greater part of the materials which form these rocks of the second series, was derived from these mountains; for, it is indubitable that the elements of which they are formed, must have been transported from regions many miles distant from the places where they are deposited, and the mountains of Calabria from which they might have been taken, are the nearest. As to the inquiry into the origin of such impetuous and extensive torrents of water, possessing the force necessary for carrying down so large a quantity of waste material, this, we must admit, is a difficult question, and perhaps the consideration of the manner in which this transport was effected is still more difficult.

Allowing every one to conceive the events of such remote epochs,

---

* Scacchi. 'Lessons in Geology.' Naples: 1842. P. 131.

according to his own measures of probability, and trusting to future researches to reveal to us the ancient history of the earth we tread, we shall confine ourselves to a few considerations relative to the manner in which the rocks we have undertaken to treat of were deposited. We find them mostly formed of thin strata with parallel surfaces, regularly deposited one over the other, and consisting alternately of strata of limestone, clay, and sandstone, thus proving the habitual tranquillity of the water beneath which they were deposited. On the other hand, the large rolled granite rocks testify that occasionally these same waters were violently agitated. Lastly, the great number of strata, the ends of which are visible in some places, assure us that the formation of the fucoidal rocks must have occupied a long period. Gypsum is not so frequent or abundant in rocks of this series as in the other. Sometimes it is arranged in strata, or crystals of considerable size are scattered through the clay; again, it forms large deposits which do not exhibit any signs of stratification: in this case its structure is eminently crystalline. An example of its extraordinary arrangement may be observed a little more than two miles to the west of Melfi, at a place called Masseria del Gesso, and a very large deposit occurs in the territory of Marcerinaro, in the province of Catanzaro, extending more than a mile. Observing the conditions of these deposits, and reflecting that gypsum is not so generally diffused as the other rocks of the same formation, we are of opinion that its origin must have depended on particular causes, probably of the same nature as those which in our days, on a smaller scale, are generating gypsum in the valley of Ansanto, in the province of Avellino. The largest crystals of gypsum formed in argil, are found close to the village of S. Potito, south-east of Piedimonte di Alife. It is not scarce in Terra di Lavoro, being found distinctly stratified near Mola di Gaeta, Casanova, Torrecuso, and in some other places. In the province of Cosenza, strata may be seen in the clay on the right bank of the torrent Pantusa, between Cerisano and Marano, and near the salt pits of Altamonte. In the last region gypsum forms a part of the immense deposits of Rock salt, of which (as it is foreign to the aim of this work to treat of them more particularly) we shall only say, that in our opinion they belong to the fucoidal rocks, and their origin is analogous to that of gypsum. The fossil characteristics of the Macigno formation belong almost exclusively to the vegetable kingdom; different species of fucus are the most remarkable, so that in some places immense numbers of impressions have been found. We met with examples of this kind in the grey marl and limestone in the vicinity of Alberona, in Capitanata, or in the red schistose marl on the banks of the little river which runs at the foot of the hill of Melfi, on the north-west side; and in the same red marl heaped together, impressions of the fucus, *Colle delle macine*, are found near Lama in Abruzzo Citra. We frequently find, both in the marl and limestone, branched cylindrical concretions, more or less broken, sometimes more than six decimeters in length and easily separated from the rock which contains them. We cannot doubt that they are formed from plants, and they ought probably to be considered fucoids. Small deposits of lignite are also frequent, among which it suffices to mention that of the Vallone della Salla, near Pagliari, to the south of Benevento, in which we have found the stems, leaves, and seeds of carbonized plants in good preservation.* It would certainly be of great benefit to science if the species of these plants were precisely defined; we are not aware whether any one has as yet directed attention to this matter, or has published the results of his inquiries. We have not time for it, nor could difficult inquiries of this nature find a place here. As to fossil animals, to repeat our former statement, if there are any, they are very rare. In some limestone strata in the vicinity of Gaeta, which probably belong to this series of rocks, we saw some very distinct impressions of Pecten, which did not appear to us to belong to any of the living species of our seas. And in the limestone near Madonna di Macera, north of Melfi, we found a few fragments of marine shells, but we could not determine with certainty to what genera they belonged. From what has been now said, the difficulty of referring our fucoidal rocks to any formation of a determined epoch must be very evident. However, as they are subsequent to the Apennine limestone, and more ancient than the supercretaceous deposits called sub-Apennine, the

* Breislak, notes this deposit in the 'Physical Topography of Campania.' Florence: 1798 Pp. 63, 64.

question to be decided is, whether they belong to the last of the cretaceous groups, or to the first of the supercretaceous. We do not know whether the question can be further settled, nor do we consider it of sufficient importance to repay the trouble of a closer definition.

Some of these rocks have been confused with the supercretaceous deposits, others have been referred to the cretaceous or (*giurassico*) Jurassic period. In our opinion they all belong to the same formation, having a similar mineralogical composition, the same paleontological characteristics, and a not discordant arrangement of strata. We consider them, then, as distinct from the real supercretaceous deposits (which they rather resemble in mineralogical character), not only on account of the want of fossil animals, but, what is more important, on account of the disagreement of their strata with that of the marls, and sub-Apennine shelly sands. We have already seen that the difference between them and the cretaceous deposits, or at least those of the Apennine limestone, is still more striking, and therefore we are of opinion that the fucoidal rocks form a distinct system.

Third Series.—SUB-APENNINE ROCKS.

The rocks of this series most frequently consist of marly clays, sandstone, limestones, and a particular conglomerate of large pebbles. The arrangement of these rocks is not so regular as those of the preceding series, the strata are not so distinct, and they are always found horizontal, or but slightly inclined to the horizon, so that they do not appear to have been disturbed from the primitive position in which they were deposited. The topographical configuration of these rocks, which are often of such little thickness that we might even call them superficial, has no distinctive character. For example, lying over the cretaceous limestone of the *Murge*, they merely render the plain more uniform or level, which but for them would have had greater inequalities. In the midst of the Apennines, or at the foot of these mountains, they form hills with a gentle descent somewhat level on the top. And if in any rare instance, as in that of the eminence upon which the city of Ariano is built, they have a more elevated and developed form, it appears to have been occasioned by the ancient topographical conditions of the place having been changed, a great part of the rocks having been carried elsewhere, while they continued with those that remained. The sub-Apennine limestone is usually tufaceous in appearance, very friable, and almost entirely formed of minute fragments of zoophytes and marine shells, many specimens of which are found enclosed in a complete state of preservation in the rock. This rock, which is most abundant in the province of Bari, we have never found in the heart of the Apennines. Sometimes it is not so tenacious, and not so rich in fossils. The sandstones (*arenarie*), are generally friable, and one might rather say were deposits of sand; thus they are easily distinguished from the compact sandstone of the preceding formation, called Macigno. They often enclose small pebbles of different kinds, which, gradually increasing in size and number, at last form a conglomerate of large pebbles, of which there are amazing deposits. The greater number of these pebbles are formed of limestone, often marly, of Piromaco which sometimes changes into Diaspore, and of very compact sandstone. There are some pebbles of granite and other crystalline rocks, which we can easily understand came from the Macigno of the preceding formation, which, as we have seen, sometimes contains them in great abundance. From the other rocks of the same formation, or from the Apennine limestone, they have undoubtedly obtained other kinds of pebbles. We may mention those of quartz and Piromaco, which often still preserve between two opposite surfaces the limestone in which the Piromaco had been imbedded in the form of little strata, precisely as we find it in rocks of the first series. Deposits of pure sand with small pebbles are met with everywhere; but the large pebbly conglomerate is only found in mountainous regions, or near them, whilst the Calcareous Tufa exclusively covers the plains. The argil then, which is always more or less marly, is habitually of a sky-coloured grey, and, owing to its plastic qualities, is much better adapted for pottery than the fucoidal argil. In Calabria, where there are extensive mountains of rock crystal, the sub-Apennine deposits occasionally manifest particular characteristics, owing to the mineralogical elements of the neighbouring mountains which

enter into their composition. One of the most beautiful examples
is furnished by certain conchiliferous breccia, in the vicinity of
Cosenza, with fragments of granite, and a great quantity of mica,
which at first sight one might say was formed of granite, and which
might lead the unskilful to think that they had found granite filled
with marine shells like sedimentary rocks. The limestone of this
series, whether tufaceous or compact, is more abundant beneath, or in
the more ancient deposit; whilst the argils and sands usually occupy
the upper position whenever they are found with the limestone.
As to the conglomerate of large pebbles, it may be considered the
most recent sub-Apennine deposit, as it is never found beneath any
other kind of neptunian rock of this series. It is frequently found
lying on rocks of the preceding series; the conglomerate on which
Ripacandida is built, and that on which the city of Lavello is
founded, furnish us with examples in the vicinity of the Vulture;
the latter deposit of conglomerate extends level for many miles,
from the tavern of Rendina, to the northern district of the territory
of Lavello. The southern declivities of the Vulture which are
included in the name Monticchio, are formed of deep deposits of
the same conglomerate, which will by-and-by claim our attention.
The cities of Venosa and Carbonara are also built on extensive
deposits of large pebble conglomerate, the arrangement of which
differs so much from that of the preceding instances, because it lies
on conchiliferous sub-Apennine marl, which is distinctly visible
near La Fontana de' trenta Angeli, little more than two miles N.N.E.
of Venosa. The immense quantity of pebbles which are frequently
found heaped together in the heart of our Apennines is an evident
proof of the great diluvial action to which these regions were
subjected after the deposition of the supercretaceous rocks.
Abandoning the inquiry into the unknown cause of this catas-
trophe, we may, with better hope of success, seek to discover
whether it preceded the emergence of Southern Italy from the sea,
or whether it was subsequent to it. Of these opinions, the
second is perhaps the more likely to be true, as no marine fossils
have ever been found in the conglomerate of which we have
spoken; and although we have never been so fortunate as to meet
with any land animal remains, yet in the Mineralogical Museum

of the Royal University of Naples, there are some elephants' tusks (*difesi elefante*), found near Chiaromonte in Basilicata, and an upper jaw with the molar teeth, belonging to the same class of quadrupeds, which was found last year near Chieti, to which fossils some pebbles are adhering, which lead us to presume that they had been dug out of the conglomerate.

Sub-Apennine paleontology is distinguished by possessing many species yet existing in the Mediterranean Sea. The subject has been treated by Brocchi and Philippi, and several other notices have been published by Neapolitan writers. Meanwhile the question whether deposits are found with us, which belong to the lower supercretaceous rocks, otherwise called Eocene, in which fossils of species analogous to the existing species are much less frequent, needs further investigation; and as we cannot conclusively determine it, it suffices to remark that the fossils of the supercretaceous rocks of Pizzo, in the province of Catanzaro, mostly belong to lost species, and that the other deposits of Gargano manifest the same condition still more strikingly.

PART II.

THE NARRATIVE AND FIRST DEDUCTIONS.

CHAPTER I.

THE REGION OF OBSERVATION—ITS SEISMIC HISTORY—OBSERVATIONS AT AND AROUND NAPLES.

THE seismic region to which this Report refers reaches in its most extended sense from Rome to Otranto, in a west and east direction, and from Gargano to Reggio in that north and south; since within the whole of this surface—in fact, over the whole of the peninsula south of the parallel of 42°—was the earthquake of the 16th December, 1857, more or less perceptible. In the more restricted sense, however, in which the seismic area is limited by the effects of the shocks having been forced upon the attention of the inhabitants, and left tangible traces of their advent, and thus determined the scope of the writer's observations and inquiries, it may be said loosely, to be bounded by a line stretching eastward from Sermoneta, at the head of the Pontine Marshes, to Foggia in Capitanata, and thence to the Adriatic; and comprehending all south of that, excepting the peninsula of Otranto, east of a line between Monopoli and Taranto and the peninsula of Calabria Ultra, south of a line from Cape Suvero to Cape Colonna, thus embracing the surface between lat. 39° and 41° 30′ and from long. E. 10° 30′ to long. E. 15° or more than 200 English miles by above 160.

This region forms part of the great seismic zone of the Mediterranean, stretching westward along the Atlas chain to the Azores, and eastward through Dalmatia, Albania, the Greek Archipelago, to Smyrna and Constantinople, and into Asia Minor, whence it bifurcates into the separate systems of Syria on the south and of the Caucasus on the north, and is one of almost constant disturbance.

Its evil celebrity has become popular through the terrible earthquake of Calabria in 1783; but the frequency and extent of its earthquakes are but little known generally.

Besides innumerable minor shocks at various points, and extending to greater or less areas, earthquakes are on record as having occurred within it in the following series of years, all of which have been of power sufficient to overthrow towns and destroy numbers of human beings, namely, in A. D.

1181	1509	1602	1654	1702	1770	1826
1230	1523	1609	1659	1703	1777	1832
1282	1537	1614	1660	1706	1782	1835
1343	1544	1617	1662	1731	1783	1836
1349	1549	1620	1670	1732	1784	1841
1446	1550	1623	1683	1743	1789	1847
1448	1551	1626	1685	1744	1805	1851
1450	1559	1638	1687	1746	1806	1854
1454	1561	1640	1688	1753	1807	1856
1456	1594	1644	1693	1756	1812	1857
1460	1596	1646	1694	1759	1814	
1486	1599	1652	1697	1767	1818	

or 82 great earthquakes since the commencement of the twelfth century.

Most of these are recorded in the Catalogues of Perrey, and in the General Earthquake Catalogue of the British Association; but the particulars of many are only to be found in the works of local authors, such as Colosimo, Onofrio, Grimaldi, Lombardi, Battista, Don Arabia, Colletta, &c., which are scarce and not to be consulted collectively out of Naples.

As observed in all habitual seismic regions, the great shock of December, 1857, was preluded by several minor ones. In 1856 shocks are recorded by Perrey ('Catal. Ann.') in January, March, May, August, October, and November, some of which embraced areas extending simultaneously to Naples, Melfi, and Cosenza; and fully an equal number are said to have marked the succeeding year prior to December: the most important in either year, was felt chiefly at and around Salerno on the 12th Oct. 1856, and seems to have escaped the usually extensive information of Perrey. The following notice of it occurred in the 'Times' of 16th Oct. 1856—

"EARTHQUAKE AT SORRENTO, Oct. 12.—The following account of an earthquake at Sorrento is given by a correspondent:—'A few hours ago we experienced two shocks of earthquake more severe than have been felt in these regions for several years. A few minutes after two o'clock A. M. I was awakened by a sensation as if my bed were about to slide out of the window in front of me. From previous experience I instantly became aware of what was taking place, and lost no time in collecting my family in the doorways of the sleeping-rooms, which are supported by very thick walls. The oscillations continued in rhythmical intervals of three seconds until I had counted four of them. After a state of quiescence—it might have been three minutes—the house began to reel confusedly, and then composed itself into another series of pendulum-like oscillations, in a direction from east to west, more prolonged than the former. I noticed that I could count, with moderate haste, three for the advance movement, and three for the return. These were repeated five times, and accompanied by a rushing noise, as of a brewing storm, and an underground rumbling like distant thunder. Indoors the sounds resembled the straining timbers of a ship in a gale. The moon was shining serenely, and the column of white vapour was issuing from the summit of Vesuvius calmly as usual, but the hurried prayers and sobbing ejaculations of the peasants in a neighbouring podere (farmstead), and the frightened baying of the watchdogs in the orange gardens, gave evidence of the terror which had just passed over the plain of Sorrento. By some the visitation had been expected. The weather had been very sultry for several days, and a peculiarly dense and ill-smelling fog had obscured the bay. The general alarm was very great, and most of the inhabitants of Sorrento rushed into the streets and open spaces. I have not heard that damage was done to any of the houses.' "

The very day preceding this, 15th Dec. 1857, a severe shock was felt at Rhodes, indicating that very distant points in the same seismic zone were then in agitation.

In recording from my note-books, &c., the facts ascertained, I shall now, for convenience, adopt the personal pronoun.

The preliminary inquiry of a day or two at Naples decided the general plan of investigation to be pursued. I proposed to myself to search the heart of the shaken country from the westward, then to pursue a southern track until I should have nearly reached its confines in that direction, and then (by another road, and further eastward if possible,) retrace my steps, and pursue a northern course until I reached its northern confines.

I reserved for the results of so much of my examination, the determination of whether to pursue a direct eastern course, then from the middle region, or to stretch farther north, and then observe the effects of the transverse chain of the Apennines upon the main earth-wave.

Much interest attaching to the question of a permanent rise or fall in the level of the land, at or after the shock, I made such observations as were practicable at and around Naples; but none of these proving decisive, I resolved, before turning to the eastward along the course of the Salaris, to go a little further south, and examine the coast and river mouths between Naples and the mouth of the Salaris and down to Pæstum for evidences of any such movement.

The narrative, therefore, will be given in the order nearly of the places visited, condensing together observations, &c., made at different times when the same places were visited twice.

NAPLES, CITY.—The following is the official notice of the shock, translated from the 'Giornale Reale' of 17th Dec. 1857:—" We have received the following letter from the director of the Astronomical Observatory at Capodimonte—'Sir—I hasten to inform you that last night at ten minutes after ten o'clock (tempo meridiano) a shock of earthquake occurred which lasted about four or five seconds; two minutes afterwards another shock of much greater intensity occurred, which lasted about twenty-five seconds. They were both undulatory, and proceeded from the south to the north. The severity of the second shock was apparent from the fact, that two pendulum clocks belonging to this Observatory which oscillated in the plane of the prime vertical, were stopped, (three others, however, were unaffected). The foundation of the tower in which our equatorial instrument is placed also sustained injury. We were also sensible of three successive but slight shocks, at three and at five o'clock in the morning (*i. e.* of the 17th.'") (See Appendix, No. 3.)

On visiting the Observatory I was unable to converse with the astronomer, Signor de Gasparis, who was unwell, but was shown over the establishment and my inquiries answered, by Signor Nobili, filio.

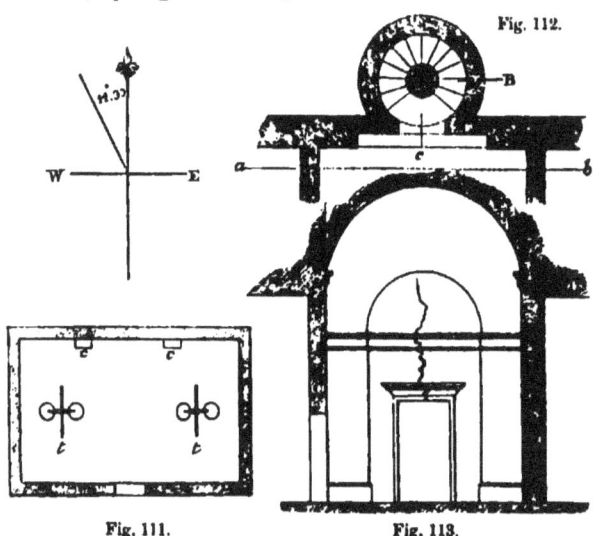

Fig. 111.   Fig. 113.

In the transit room (Fig. 111) (*t t* being the two

instruments), are two clocks, *c c*, whose pendulums vibrated in the plane of the prime vertical, and showing sidereal time. These were both stopped, but, according to Signor de Gasparis, at different moments, and each at an unequal period after the shock, owing to their structure, so that nothing could be concluded as to the precise moment of the first shock from them. I could not ascertain upon what precise data the moment stated in the Giornale as above, for the occurrence of the shock was based, and from other facts entertain some doubts as to its precision.

I found by measurements that a moment in the line of the meridian, and therefore transverse to the plane of vibration of less than 0·5 inch would have been sufficient to have stopped either of these clocks, unless the contact with the case and pendulum so produced, had been instantly removed by a movement in the opposite direction, and before time were given to destroy by friction the momentum of the pendulum.

In the Salle Centrale, which is also the library of the Observatory, and leads by a winding stone staircase at one end, to the top of the tower where the equatorial is fixed, is a third clock, showing Naples mean time, whose pendulum, an extremely heavy one, oscillates in the plane of the meridian, which was not stopped. A movement of the pendulum bob of 0·625 inch transverse to the plane of oscillation would have stopped this clock.

These clocks are not screwed to the walls, and neither they nor any of the other instruments had suffered damage or derangement.

Two chronometers lent by the Lords of the Admiralty, and brought with me from England, going Greenwich mean time, were compared with the time at the Observatory, and one was found to have been slightly deranged by the railway journey. I had deemed it probable, that by the co-operation of some Neapolitan savant, I should be able to get measures of horizontal surface transit velocity of the earth-wave, in some of the slight shocks said to be still continuing. The uncertainty of their recurrence and difficulties as to finding any Neapolitan co-observer, rendered this impracticable. The chronometers were, however, of much service to me in the interior of the country, one of them having been adjusted to Naples mean solar time before I started.

The Salle Centrale has its length in the direction of the meridian. At one end a doorway, *c* (Fig. 112) leads to a stone winding staircase, descending one deep story, and ascending to the equatorial, which is thus placed on the top of a cylindrical tower, formed of a central solid cylinder of masonry of about 6·5 feet diameter, the steps about 4 feet wide, and the outer cylindrical wall of about 3 feet in thickness; the total height from the ground to the floor of the equatorial being about 70 feet.

Fig. 113 is a section across the Salle Centrale at *a b* (Fig. 112), showing the interior elevation of the end next this tower. From the centre of the lintel of the doorway at *c*, a nearly vertical fissure, open 0·20 inch at bottom, extends upwards, becoming evanescent at about twelve feet, and its plane is in that of the meridian. Its continuation downwards can be traced from the centre of the sill of the doorway also.

A second fissure at B, occurs right through the outer wall,

at right angles to the former, or east and west, and extends about ten feet up and down, (commencing where the wall had been weakened by an aperture now built up) and vertical- -width 0·15. It is higher up the tower by five feet, than the fissure C at its mid length. The inertia of the central core of masonry here is enormous in relation to its base; and to that, no doubt, is due these fissures, the only two formed in the whole building, which is solidly and well built of rubble and ashlar masonry.

The fissure C appears to have been produced by the spiral lapping of the staircase round the central column (through which the push of the latter was transmitted to the cylindrical shell) having prevented its vibration as a simple pendulum in the plane of the shock, or near it, and induced a movement of conical vibration. The movement indicated is one nearly from south to north by compass.

Several other buildings in and directly around Naples were fissured, but none were thrown down. Amongst those which I examined were Messrs. Turners' bank, in St. Lucia, the Tribunale, and several palazzi, amongst the latter the Palazzo Lieti, in the Toledo. In no case, however, could I find that the fissures had been *originated* by the shock of 17th December: they were all pre-existent, and due chiefly to settlements, but had all been slightly enlarged by the shock.

The derangement at the Palazzo Lieti was so considerable as to demand prompt measures to prevent the fall of one wall of the interior court, by building up solidly, a huge arched porte-cochère that had yawned beneath it, and was about twenty-four feet span, with a new wall and smaller arch as in dotted lines Fig. 114. The building, consisting of four lofty storeys, and nearly eighty feet in height, I found had been fissured from settlements for a

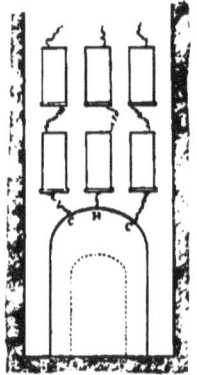
Fig. 14.

length of time; but the shock had been sufficient to shake downwards the central mass of the wall between c and c, and to widen all the old fissures, which were now three-quarters of an inch wide—those c and c widest at top, B widest at bottom—evidencing clearly the nature of their production.

This, I found, was considered the most formidable example of injury to buildings occurring in or around Naples.

In no case, except at the Observatory, was I able to remark an original fissure in any well-built and sound structure.

The actual range of movement at Naples must have been small and far from violent. The amount of alarm produced generally by the shock was, however, sufficiently great to cause almost the whole population of the city to spend the remainder of the night of the 16th December in the open air, in carriages, around large fires in the streets, &c. The principal source of alarm described by most persons was from the creaking and straining noises of the timber work of the heavy floors and roofs, and the rattling of the windows and doors. A large portion of the population spent the succeeding night of the 17th December also in the open air, or in parading the public places. It was manifest, however, that much of this on both nights arose from the excitement and newsmongering tendencies of the people, who made a sort of "festa" of the occasion, and but little from actual terror except at the first moment.

From some persons of observation and discretion, I collected, their own perceptions of the phenomena.

A young English lady, residing in Santa Lucia, of much intelligence and observation, was at tea with some friends, sitting round a table whose length was nearly E. and W. by compass. Her attention was first arrested by a transverse movement of the table sliding back and forwards about an inch each way upon the waxed tiles of the floor. This she at first thought arose, from some of those who sat with her, but on casting her eyes upwards, on hearing the floor above creaking, she saw that a lamp suspended from the centre of the ceiling was oscillating also. Earthquake, which she had experienced elsewhere, then occurred to her, and she noticed carefully both the direction in which the lamp swung and the arc of its oscillation. She set the lamp itself again swinging for me, above the same table, in as precisely the same direction and to the same extent as possible. The direction I found to be 8° 0′ E. of N. by compass, and the summit of Vesuvius bears 110° E. of N. from the front window of the room (which was on the second floor from the ground). The chord of the arc of vibration was $10\frac{1}{2}$ inches, and the lamp makes thirty double oscillations per minute by the watch; it weighs about 12 lbs.

Her sensation of the shock, was of a small, rapid, recurrent, movement, forward and back, perfectly horizontal, without any undulating motion; then a cessation for two or three minutes (as estimated), and again a renewal of the same motions; after which all was quiet.

The late Dr. Lardner was residing with his family at the Hôtel des Iles Britanniques, in Chiaja, and was at the time in a "salon" upon the third floor. His impressions were of a larger amount of oscillation, and with more or less of undulation. He has recorded them (with, perhaps, a little excess of colouring) in a letter published in the *Times* of 29th December, 1857, under date 19th December. He obligingly went with me to the "salon" at the hotel, wherein we found a large and ponderous chandelier hanging, of which he had observed the swing on the night of shock. He set it again in movement in the same direction and to the same extent.

This chandelier weighed 190 lbs. avoirdupois, hung (to the lowest point) 8 feet 9 inches from the ceiling, and by trial made $20\frac{1}{2}$ double oscillations per minute.

According to Dr. Lardner, it commenced to swing in an arc of about 24 inches chord, and in one plane, the azimuth of which I found to bear 13° 0′ E. of N., the Point of Pausillipo bearing 130° W. of N. from the front windows of the room. This vibration rapidly became elliptical, the major axis diminishing from 24 inches until it became about 12 inches; when the lamp continuing to vibrate as an ellipto-conic pendulum, was stopped by Dr. L., as he stated, both to appease the alarm of his family and to enable him to observe the effect of a renewed shock. The time by his watch was 10^h 15′ Naples mean time, but he could not guarantee that the watch was perfectly right, though a good one.

His sensation of the first movement was of a short, jarring, horizontal oscillation, that made all doors and windows rattle, and the floors and furniture creak. This ceased, and after an interval that seemed but a few seconds was renewed with greater violence, and, he thought, with a distinctly undulatory movement, "like that in the cabin of a small vessel in a very short chopping sea." It was sufficient to demand a certain amount of attention and effort, on the part of those standing up, to maintain their equilibrium.

Signor Guiscardi, a highly intelligent observer, educated as an architect and civil engineer, and well acquainted with physical science generally, had just retired to bed, when his attention was aroused by the first movement, which he describes by a little diagram, as simply a short sharp, jerking movement forward and back, thus:—

within narrow limits, and lasting, as he supposes, about five to seven seconds; then a total pause of some seconds, and then the former movement recommenced with rather more violence, and in this sort of order—

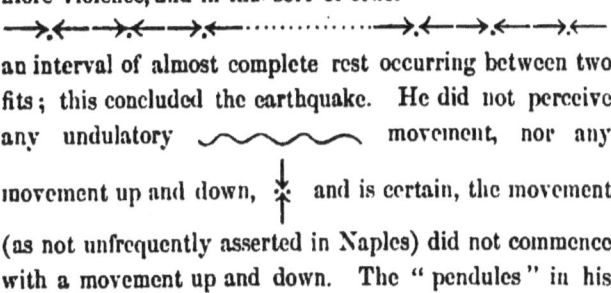

an interval of almost complete rest occurring between two fits; this concluded the earthquake. He did not perceive any undulatory movement, nor any movement up and down, and is certain, the movement (as not unfrequently asserted in Naples) did not commence with a movement up and down. The "pendules" in his rooms, having a general E. and W. plane of vibration, were stopped at $10^h 10'$ Naples time, but he cannot guarantee their accuracy as to time.

I was not enabled to gain any additional facts of importance, from conversation with Signor Capocci, ex-professor of astronomy, or with Signor Palmieri, professor of physics. Both agreed that the direction of wave movement was from S. to N., with more or less of an eastern or western swerve from the magnetic meridian.

There was no material alteration, either in inclination or declination of the magnetometers noticed; but Professor Palmieri's views are, that every eruption of Vesuvius and Etna, and probably every earthquake, is accompanied by great electrical disturbance, which, he supposes, may affect the magnet. His seismometer at the Observatory upon Vesuvius was affected by the shock. The magnetic declination is very variable, both in short periods of time and for adjacent localities in and about Naples, and he thinks continually alters with the state of Vesuvius. I myself ascertained the declination in St. Lucia to be only 9° west in one spot; but blocks of lava used in building, pavement, &c., all more or less magnetic, make such observations very uncertain. The mean declination, however, for Naples, I obtained from the Observatory = 14° 30′ west in February, 1858; and this agrees with the monthly printed determinations of the Royal Marine Observatory at Naples.

The whole of these five observers above mentioned agreed, that the shock at Naples was not attended with any noise whatever, either preceding, during, or succeeding the movement.

One gentleman only in Naples described to me sensations of sickness felt by him during the shock, which he first perceived while playing cards. In his case my impression was, that the affection was due to nervous excitement and alarm only. Conversation generally with persons of all classes in Naples, only tended to increase on my part the caution necessary in attempting to found any conclusion upon statements of physical facts, so exaggerated and often inconsistent, as those in common circulation, though unaccompanied by intentional deception.

Signor Fiodo, Vico Baglievo, Strada Toledo, chronometer-maker to the Neapolitan Marine (who executed the needful repair and new rating to one of my English chronometers), I found a man of great intelligence, discretion in observation, and accuracy of thought, and from him I derived some of the most useful facts obtained at Naples.

In his "atelier" he has a regulator clock, with a heavy gridiron pendulum vibrating seconds, and, as I found, oscillating in an azimuth 20° E. of N. by compass. Resting upon the bottom of the clock-case, which is of polished chestnut, he had long placed a small steel anvil or parallelopiped of the exact dimensions shown and figured in Fig. 115, five sides of which are smooth, but black as when forged, and the sixth polished. This stood on edge, the polished side being next the pendulum (and behind it as one faced the clock), and the plane of this side, parallel with that of the oscillation—the polished surface being by measurement exactly 0·276 inch horizontally from the adjacent side of the screw at the bottom of the pendulum bob. The chord of oscillation of the pendulum measured at the screw was = 1·87 inch, and less than the parallel dimension of the steel block. A small amount of transverse movement, therefore, would be sufficient at any time to stop the clock by bringing the screw of the pendulum into contact with the face of the steel block.

On the morning of the 17th December, 1857, Signor Fiodo found this clock stopped, and the screw of the pendulum in contact with the south end of the steel block, which had been shifted from its place by the momentum of the pendulum, as in Fig. 115.

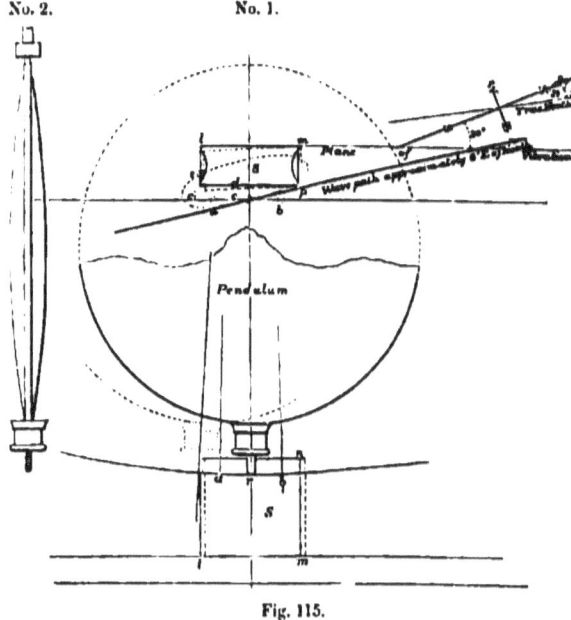

Fig. 115.

I shall recur to the inferences derivable from this in Part III. It is sufficient here to remark that the direction of wave movement indicated is one approximately 6° E. of N. The clock was going (tempo meridiano) or solar time for Naples, and was on the evening of the 16th December true to time within an error of 0·5 second. It had been stopped at 10ʰ 13′ 26″ P.M., which was therefore the time of the first shock at Naples, within less than half a second the error of the clock (slow), + the minute fraction of time due to the increased semi-arc of vibration.

There were several other clocks, some with pendulums,

but making half and quarter seconds beat, in Signor Fiodo's establishment, oscillating in directions approximating to the meridian, and to the prime vertical, but none were stopped, and on examination I found them not circumstanced so as to have been so.

I verified with care all the dimensions and particulars of the regulator that was stopped, and have not the smallest doubt either of the good faith upon which the facts taken from Signor Fiodo rest, or of the exactness of the conditions as observed by him, and the dimensions, &c., as taken by myself.

On passing into the Naples end of the tunnel (or grotto) at Pausillipo, I remarked several fine, keenly-drawn lines of nearly vertical fissures, in the perpendicular banks of yellow tufa at the right-hand, or S.E. side of the entrance, which appeared recent. The light, however, was not sufficient to enable me to decide. I therefore returned early the following morning, and by clear sunlight made a minute examination of those cracks (the occurrence of which no one had remarked as far as I could learn). I satisfied myself that they were very recent, that they were not due to any settlement or alteration by gravity alone, of the banks, nor due to any artificial work, or excavation.

The keenness of their external lips or edges, the absence of dust, cobwebs, or insect or vegetable life, within or across them, and their narrow and uniform breadth of opening about 0·2 inch, their general parallelism, and, above all, their direction in azimuth, with relation to the form and direction of face, of the bank, and their verticality, convinced me that they had been produced by the shock of the 16th December, and were due to the inertia of the enormous bank of tufa through which the tunnel has been cut. I found their direction to be such, as indicated distinctly a wave-path and direction, of between N. 20° W. and N. 38° W. The last extreme appeared to me doubtful, and as only derived from one cleft. I am disposed to adopt the former azimuth only.

Upon the whole, the indications of wave-path at Naples are meagre, though not indistinct, nor discordant. They vary between the limits of N. 13° E. and N. 20° W., or, omitting Pausillipo wholly, vary between N. 6° E. and N. 13° E., and comparing *all* the indications, seem to give a resultant path of, N. 6° to 8° E. as the most trustworthy.

This appeared to point to a focus somewhere at sea, beneath the gulfs of Salerno, or Polycastro, a first impression that became not a little puzzling, when brought into contact with the facts, as they developed themselves in the interior provinces, and at first, for a day or two, almost caused me to despair of being able to trace out the true focus at all, the fresh evidence as collected appearing to be quite conflicting; and it was not until after I had found reiterated proofs of an inland focus, that could not connect itself directly with Naples, that the solution of the difficulty began to appear, in showing the shock at Naples city, to have been merely a reflected and refracted one.

CHAPTER II.

PERMANENT CHANGES OF LEVEL ACCOMPANYING EARTHQUAKE—THE THEORIES OF SERAPIS.

So much interest attaches to precise observations, as to permanent change of level of the land, occurring at the same time with earthquakes; and this object having been urged upon my attention, by my friend Sir Charles Lyell, before I left England; I therefore gave the question of whether any such change had attended this earthquake very careful investigation, and I may say, have examined, as to it, the whole coast at various points, from north of Pozzuoli to Pæstum. I found the almost universal opinion at Naples was, that an elevation of some inches around the whole bay, varying at different points, had taken place, and the circumstantiality, with which intelligent persons residing upon the shore, pointed to apparent proofs of their impression, demanded much caution. Professors Capocci and Scacchi, with Signor Guiscardi, doubted the existence of any change of level, but could give no facts either way. All the evidence presented to me, was based upon references to assumed changes of tidal level.

The English lady at St. Lucia, before referred to, pointed

to a sloping quay bench (Fig. 116) opposite her windows. She had always remarked, that, at high water, the tide covered to the point c, or an inch or two above it, prior to

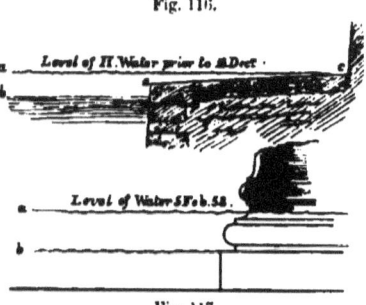

Fig. 116.

Fig. 117.

the 16th December; but since that, the high-water level had been permanently about five inches beneath the arris of the quay at e, giving a difference in level of from nine to twelve inches. To test this I examined the water level daily at the hour nearly of highest tide, and for four days found the highest tide-mark as at b; but on the next occasion of observation it was not only at c, but some inches above it. The difference was simply due to the off or on shore wind.

It would be tedious to record several other observations round the bay of like character.

On visiting the Temple of Serapis, at Pozzuoli, where the notoriety it had already acquired on this point, and the daily attention given to it, presented the best chance of decisive indication, no evidence whatever could be found of change of level. The "gardien" of the place, however, on being questioned as to whether *he* had observed any change of level, at once directed our attention to the base of one of the worm-eaten columns, and stoutly affirmed that the level of the water which was then standing at *a* (Fig. 117), had, directly after the shock of December, fallen to *b*, equivalent to a rise of the temple of 7 inches, but that, since that time, the water had gradually returned to its former level, *i. e.* the land had sunk again.

He denied that the difference could be due to variability in the sea level. The utmost limits of disturbance by wind or tide within the sheltered valley of the ruins being, according to his stated experience, far within 7 inches.

I could not find, that any man of science in Naples, had ascertained what these limits of aqueous disturbance were, and on my return to the city (from the interior), I took the occasion of a severe gale of wind in shore—the "Garbino," from the S.W.—and at the presumed time of high water, to visit the temple again, in company with Signor Guiscardi, when I found the water rather above the level of the sill of the entrance iron gate, and fully 22 inches above the level of the 5th February, and it had been nearly 3 inches higher about two hours previously.

It is obvious, therefore, that any deduction whatsoever as to levels, whether of elevation or of depression, based upon the tidal level of the Mediterranean on this coast, cannot be depended upon, within the limits of 18 inches or 2 feet at the very least; and several of the speculations as to minute oscillations of level of the Temple of Serapis so based must henceforward be received with doubt.

Impressed with this fact, in which I found that Professor Capocci and Signor Guiscardi coincided with me, and with the extreme value to physical science, of possessing, in this instable region, some definite and unimpeachable standard of level, I addressed a formal letter, upon my return to Naples from the interior, to the government of his Majesty, the late king, suggesting the importance, of having an accurate line of levelling run through to Naples, from the sill of the front door of St. Peter's, at Rome, which may be presumed at present as the best, if not an invariable datum point, and the difference of level marked upon bench marks at and around Naples. (See Appendix No. 4.) The work could be performed with ease and little cost by the officers of the " Ponte e Strade," going along the high road between the capitals. I regret to say, however, that it was intimated to me, at the Ministry of the Interior, that this despotic government objected to entertain suggestions from foreigners, even as to matters of science; and the work, which could then have been accomplished with facility, in connection with certain railway surveys in progress, remains, and is likely to remain unperformed.

While the limits of error as to levels deduced from the sea, affect all *minute* questions of rise and. fall of Serapis, they do not touch the great change of level, as evidenced by the celebrated columns; but they appear to me sufficient to destroy the force of the conclusions of Niccolini and others, as to oscillatory changes of level of small extent.

The evidence of elevation, of the whole building since its original construction appears to me irrefragable; but not so that upon which the supposition of its subsidence first, after its erection, and previous to its elevation are based. The argument for subsidence, rests upon the improbability that the level of the floor of the building was originally designed and constructed, below that of the mean tide of the Mediterranean. Now it appears to me that the probability runs just the other way. Archæologists appear to have settled, that the so-called Temple of Serapis was not a temple at all, but a public bath, a conclusion that forces itself upon the mind of any untheoretical observer of the general architectural structure of the place. If a bath, nothing is so probable, as that its level should have been fixed with reference to the sea, such that sea-water would run in, or command the baths, in a place where there appears to have been no fresh water except that of the thermal spring. The possible objection to this, that there would then be no drainage for the waste water of the baths is met by the fact, that the dry and porous subsoil, consisting of 12 to 20 feet of tufa, lapilli, and scoriæ, would soak away any amount of water, if simply discharged into a pit sunk in it, below the level of the baths, a method of drainage actually practised from a remote age to the present day. A considerable district of Paris at present discharges the whole of its sewage into such a "puit d'absorption."

The land at the existing level of the terrace called La Starza, upon which the temple was built, is in rapid and constant process of marine degradation at present; so much so, that unless artificial means be soon taken to prevent its inroads, the sea will in another half-century probably, have swept away the whole temple (so called).

It therefore was probably very much more inland when first constructed, and was probably built either in some natural depression, of 10 or 12 feet below the sea level, or in one excavated to that depth, by a race whose *burrowing* tendencies are revealed by many of their buildings, in all directions around. If much inland, there was doubtless a sufficient mass, though of porous material, between it and the sea, to be water-tight; but if, as more and more of this became removed, the sea-water percolated the bank universally, at the seaward side, it could no longer be kept out from the building, and the place would have been abandoned as untenable.

The water of the sea would then stand permanently at a level with the highest line of testaceous perforations of the limestone columns, say about 20 feet above the level of the present floor, *assuming that the general level of La Starza was then about 8 feet under what it now is*, and that the floor was originally founded 12 feet below the sea level.

The channels or ducts that had before brought the seawater to the baths would also bring the young testacea, and preserve sufficient change for their healthy existence. If, subsequently, the land bearing the so-called temple upon it, were gradually elevated about 8 feet, resting at about its present level, we have sufficient to account for the phenomena observed, without having recourse to, several successive depressions and elevations.

Elevations are common, and obviously part of the established cosmos of the earth's surface, but depressions, due to subterraneous forces, appear exceptional and rare, and especially doubtful, close to volcanic vents. Land-slips and aqueous erosion, marine and of every other sort, appear the established agents for depression of surface, acting in antagonism to the former. Indeed, proofs seem wanting, of any such thing as recurrent oscillation of level, of any known tract of land within the historic period, traceable in both directions of movement, to subterraneous agency.

To the view here advanced as offering the simplest and most probable solution of the Serapis problem, it may be objected, that the adjoining ruins of the Temples of Neptune, and of the Nymphs, are some feet under water, and that the arches of the so-called Mole of Pozzuoli, are covered above the level of the springings. The levels of these two latter temples will not accord either with the presumed depression or elevation of Serapis, and may hence be made to argue as much against, as for the oscillatory view; and as to the arches and general structure, of the so-called mole now deeply immersed, I am satisfied that it never was built for a sea mole at all, and that the whole of the arches were originally built on dry land, and for other purposes. It would be a work of no small difficulty, to construct these piers and arches in the open sea-way, where their remains now stand, with all the aids that modern engineering afford: and without the diving-bell we may safely affirm that they never could have been built in open sea-water. They, further, are of dimensions and construction, that no Roman or any other architect would have adopted for a marine mole.

How, then, came they immersed as they are? It appears to me that they, and the incoherent tufaceous land, that sustains them and these temples, are now, and have long been, in gradual process of insensible land-slip downward and seaward, by the continual removal by tidal action of the loose material, from the foot of the submarine talus, which the soundings prove, to be outside them, in the roadstead, hence unequal subsidence, but always greatest where nearest the sea-shore. And this view is strongly corroborated by the fact, that *all the standing columns at Serapis lean some inches out of plumb to seaward*, and that *the whole floor of the place is waved and uneven, and with a general out-*

of-level slope to seaward also, as though the whole mass stood upon a base of loose soft material that was gradually settling and going seaward from the effects of sublittoral erosion. This seems also to be the solution, of the instances of the Roman roads, under water between Pozzuoli and Baiæ, and the Lucrine Lake.

Moreover, if Serapis had been ever depressed to the extent required, then this so called marine mole must have been equally so; but it is quite obvious to an engineering eye that were the arches, upon the piers as now standing, depressed but a few feet more, so as to receive the full stroke of the waves in storms, or the entire impulse of the moving superficial column of the sea, they would have been overthrown long ago. They only stand because they never yet were wholly under water.

The general importance of questions of permanent elevation or depression, and their intimate connection with earthquake phenomena, will, I trust, be deemed sufficient ground for this digression, upon the much-discussed Temple of Serapis.

CHAPTER III.

EXAMINATION ROUND THE COAST AS TO PERMANENT CHANGES OF LEVEL.

On my way southwards, I received an introduction from the Intendente of Salerno, to Signor Palmieri, an engineer of the Ponte e Stradi, whom I fell in with at Eboli. In conversation he stated his opinion that since the shock of December 16th, the sea level all round the west coast, has been lower, *i.e.*, the land higher than before. He sustained this, by reference to a quay or wharf wall not very long since erected by himself at Amalfi on the shore of the Gulf of Salerno. The level of the sea at half

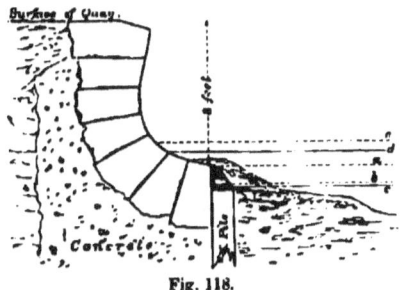

Fig. 118.

tide, he said, was always previously at *a* (Fig. 118), or coincident with the top of the timber work, at the toe of the sea wall, but then stood about one palm, or 10·38 inches English, below it at b. I examined into this on my return northwards, but the account of my observations will be best given here. On the 27th of February, 1858, at the lowest point of low water, of the afternoon tide at Amalfi, I found the sea-level to be 11 inches below the top of the cap sill or longitudinal timber, over the tops of the piling at the toe of the wharf wall, (which has a hollow parabolic curved sectional contour); that is to say, about half an inch below the half-tide level b, according to Palmieri. Some loose volcanic sand, was heaped up at the foot of the wall above the permanent gravel of the beach beneath. Within a few hundred yards, I was able to find a sheltered nook between some rocks, where I noted the usual rise of tide, by the weed marks, to be 17 to 18 inches, and that high spring tides rose occasionally about 4 inches more.

I recurred to this spot at high water of the same tide, and although having to use a lanthorn, and a little wind having sprung up, I was yet able to ascertain a rise of tide of 16 inches. Returning to the quay wall, I found the water there, too agitated for direct observation; but referred to my adjacent tide gauge, the quiescent level of high water would then have been, 5 inches above the top of the cap sill, or at d, and with an 18 inch tide, 7 inches above same; or at high springs, about 11 inches above same. So that, the half-tide level is still in reality, just about the level of the top of the cap sill at a, as fixed by Signor Palmieri before the earthquake, and no change of level of the land has taken place at this point of the coast. His erroneous conclusion, must have arisen, I presume, from his having mistaken at the time of observation, the periods of high, and of low water.

I also examined the coast carefully to the eastward of Torre del Greco, and to where the railway branches off to Castellammare, and where some of the firm lava streams from Vesuvius, have run into the sea, and afford the best natural marks as to tidal level that the bay presents, but was unable to find any evidence of recent change of level of the land.

I also examined the quays and beach at Salerno, the mouth of the river Vicentino, which falls into the Gulf of Salerno, S. E. of Monte Corvino, and the beds of the rivers Tusciano and Salaris, in the great plain of Pæstum, with the same negative result.

At Pæstum, the proprietor of the soil, who is also the appointed guardian of the antiquities of the place, was at home at his "Casone." He was perfectly familiar with every feature of the shore line, from the ancient city down to Agropoli, and had recently passed along it, but had remarked no changes since December 16th, 1857, though quite alive to the question of rise and fall of the land.

I conclude, therefore, that there is no evidence whatever of any permanent change of level of the land in connection with this earthquake, upon the west coast from Pozzuoli to Agropoli, and it is not conceivable that there should be any, upon the Adriatic coast, where the shock was only perceptible.

CHAPTER IV.

JOURNEY SOUTHWARDS—AMALFI—SALERNO—VIETRI—
LA CAVA—PLAIN OF PÆSTUM.

I now recur to my journey southwards. At Torre del Greco, Resina, and throughout the whole seaboard of the Bay, the shock was felt as sharply as at Naples; and generally over the whole plain of the Terra di Lavoro, in a direction from south to north: at Ottajana, to the southeast of Vesuvius, and close under the mountain, more than usual injury was done. The church of St. Michael was largely fissured, and that of St. Johannes Battista also. I did not visit that town, but an inhabitant, whom I met elsewhere, stated that the direction of shock was generally felt from south to north, but also seemed to come from Vesuvius, and the like facts were stated as to the village of Somma. Small and unindicating fissures were to be found, in the older and worse built houses, &c., everywhere. At Torre del Annunziata, the west façade of the church, is largely fissured in directions clearly indicating a wave-path, not far from south to north. Ancient fissures from former earthquakes at lower levels are visible in its walls.

At Castellammare, deduced from, not very well-defined fissures, the wave-path varied from 12° W. of N. to 6° 30′ E. of N.

At Sorrento Point, the direction of wave-path was described to me by several intelligent observers resident there at the time, to have been from S. to N., or very nearly so. I could not visit myself that locality.

In Pimonte, also, the façade of the Chiesa Madre was fissured, and part of the roof thrown in. At Sigliano, some houses were overthrown. At Gragnano, on the slope above Castellammare, a great many poor buildings were greatly shaken, as was also the case with all the villages, upon both the north and south sides of the mountainous peninsula, terminating with Punta della Campanella; but in the island of Capri, directly south of Naples, and but a few miles from this cape, the shock was scarcely perceived.

On the south side of this peninsula, Tramonte, Minori and Majori, were fissured, but uninstructively, from the character of the buildings. At Amalfi, the shock was alarmingly felt. The doors and windows rattled for ten or twelve seconds at each of the two shocks; but no injury occurred to any of the buildings, which are generally of a substantial and well-built character of masonry. The Padrone of the Hôtel des Capuchins, and also the chief apothecary of the place, were able to point out to me separately, the directions in which they perceived the shock; both statements closely agreed in pointing out an azimuth, which proved to be 133° W. of N.; *i.e.*, from a S. W. to N. E. direction; and several facts indicated the occurrence of an orthogonal shock here, and at Atrani. They heard no noise.

The line of coast here is nearly E. and W., and so is the face of Palmieri's quay wall.

I made inquiries here, as also at Salerno, amongst the fishermen and coasting sailors, as to whether any of them had felt the shock at sea, but could gain no intelligence of any such observer.

In the Ravina della Molini, behind the town of Amalfi, I observed some beds of ancient tufa deposited upon the precipitous sides of limestone at a considerable height above the present sea level.

Along the road between Amalfi and Salerno there are proofs of an elevation bodily of the land of from 300 to 400 feet since the formation and forcing up into a mountain range, of the great ridge of limestone that forms the peninsula. 1st. Beach gravel in wavy layers, quite similar to that on the existing shore, is found 300 feet above it at Punta d'Erchia. 2nd. Between that and Amalfi, in the limestone, the beds of which have a north and south strike, and dip slightly to the west, there are caves, the upper portions of the jaws and arches of which, some 70 or 80 feet above the existing beach, present the rounded and water-worn aspect, of long-continued action of the sea. Objects, such as the porphyry font at Amalfi, alleged to have been excavated from beneath the beach—ruins now existing below the sea level here and there, go for nothing, as along a line of coast so extremely precipitous as this, of shattered limestone, and so frequently shaken by earthquakes, whole cliffs, have doubtless frequently been shaken down, and plunged beneath the sea. The limestone all along, from the point of Capo del Tumulo, is metamorphic and altered, in its bedding and cleavage, and presents in many places, highly magnesian, and in some, trappean characters. Near Majori, fine masses of dark-brown stalactite occur, containing very large plates of calc spar. All this, with the scattered patches of tufa, on the south side of the peninsula, where they never could have come *sub dio* from Vesuvius, indicate that submarine volcanic action, was going on in these regions, before the bay of Naples was separated from that of Salerno at all, by the elevation of the great limestone ridge now between them.

At Vietri, which I visited in a violent storm of rain and wind, I could find no evidences of wave direction worthy of notice. At La Cava, the first very obvious trace of the earthquake challenged notice, in a long range of diagonal timber braces, sustaining the S.W. side of a range of house fronts, which had been thrown so as to lean outwards, bringing with them, the square piers of the old Roman-looking arcades, over which the houses are built (Fig. 119). In the latter were measurable fissures, though small; in the Casa Communale, and in the side and back walls, of some of the strange shadowy open fronted shops, that seem so identical with those of Pompeii, were a few others. From the whole I obtained three indications of wave-path — 15·30 E. of N.; S. to N.; 17 W. of N.— and also some indications of an orthogonal shock, W. to E. At the Benedictine monastery of La Trinita, a few miles from La Cava, I expected to have found much evidence of injury. It lies, in the gorge of a deep and sinuous mountain valley, of metamorphic limestone, hard and shattery, but with much diluvial covering in many places.

The buildings, sound and well constructed, of rubble ashlar chiefly, have generally escaped. There are, however, in two different places in its southern corridors, and near the great south corridor window,

fissures of considerable length, and open at widest
0·4 inch. They indicate a wave-path from the southward, and in direction 16° 15′ W. of N., and also one
105° 30′ W. of N. or orthogonal. They felt the two shocks
of 16th December severely at the monastery. Padre Morcaldi, the Archivario, was not conscious of any noise attending either of the shocks, nor had any one else in the
monastery remarked any.

The forms of the small mountain valleys in this thickly
inhabited region, are singularly winding and capricious
(Fig. 120). A shock in whatever general direction
acting here upon the houses and towers, perched on declivities, now rocky, now diluvial, and scattered here and
there and facing every point of the compass, must produce
effects in the highest degree complicated, or even unaccountable. I therefore resolved, in the first instance at
least, to waste no time by further observation within it.

Salerno, though an ancient city, is generally well built:
it lies low, along the shore of a pebbly beach, and apparently on pretty deep beds of loose material, and the land
behind it, rises gradually into mountain slopes, and recedes
into sinuous transverse valleys on limestone.

It has not suffered much, but there are abundance of large
measurable fissures. I had a lengthy conversation with
the Intendente of the province, Signor Ajosso (who was confined to bed and unable to go round the city with me).
He stated that the shock was not sufficient to throw down
furniture, or observably displace it, but that he saw it jerk
the water out of a large earthen jug, which he pointed out
in his bedroom, about 5 inches diameter of mouth, and
which had been full, within 2 inches.

VIETRI, FROM NEAR THE NEW ROAD.

Some china vases and table ornaments of common form, in his rooms were thrown upon the floor. Some of his officials lost their footing and fell, during the second shock. He heard no noise with either shock, nor has he heard it stated that there was any sound heard by any one at Salerno. He furnished me with a copy of the official list of the houses destroyed, persons killed and wounded, &c., in the several Communes of his province, and letters to the Sotto Intendenti in the various parts thereof, and a general authority, under his sign manual, to call upon the Guardia d'Urbani and Gendarmerie anywhere, if necessary to aid in my examinations.*

The walls of the Intendenzia are rather heavily fissured, especially through the window and door opes of the great stone staircase. The fissures in walls running E. and W., lean 10° to 13° from the vertical to the eastward : some are open 0·5 inch in 12 feet of length, and props and braces have been necessary.

The building is ordinal, and from three of the best fissures, I derive a wave direction, 53° W. of N.

The cathedral, a grand old structure, rich with the spoils of Præstum, brought by Roberto Guiscardi, and of sound masonry, has its axial line (as I find is not uncommon in these very old Italian churches) not quite E. and W. The axis is 23° W. of N. There were two formidable fissures, one in the apse at the N. E. side,

---

* The Intendente is very much the representativo of the ancient Roman Prætor. He possesses enormous executive power, often grossly abused under the old *régime*. The civil, military, and ecclesiastical authorities are all more or less subject to his individual will.

at *a*, Fig. 121, the other at *b*, in the western transept—both not far from vertical, and right down through window and door opes to the ground level, from the roof. These were originated, the Sacristan informed me, by small thread-like cracks (*filone*) in 1851, but were widened and lengthened now, in December 1857. *a* is now about 0·75 inch open near the top, and the roof of the apse has been sufficiently injured to require struts betwixt ceiling and floor. The dislocation of the roof here, indicates a certain amount of emergence in the wave-path (but the fissures do not indicate any distinctly), 10° or 12° at most.

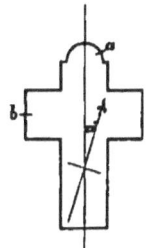

Fig. 121.

The direction I derive from them is 34° 30' W. of N. There are upon various points of the cathedral and attached buildings many slender iron crosses, the iron flat bars of about 1½ inch × ⅜ inch thick, as in Fig. 122, none of which present any signs of having been bent or twisted. They were all confined, however, by small diagonal stays of round iron, about ¼ inch diameter. In the noble old cloister court, a long stone, part of the shaft of an old column that had leaned against a wall running nearly N. and S., was overthrown, and indicated a wave-path of about 60° W. of N.

Fig. 122.

Fissures, at the Tribunale, gave a wave-path 67° W. of N. These fissures also afforded pretty decisive evidence of the co-existence here of an orthogonal shock, or one from W. to E., but of very minor intensity, as was already noticed at Amalfi, Atrani, and La Cava.

Several other churches that I entered showed no sign of injury. I was informed that the church of Saldina with its Campanile, not far from Salerno to the northward, had been seriously dislocated, more than any at Salerno. Time would not admit my diverging to it.

Throughout the whole vast plain, from Salerno to Pæstum, no visible sign of the earthquake can be found. It was felt however, sharply and with alarm, all over it, and the people very generally say it came from the eastward, in so far as their very loose expression "levante ver, ponente," may mean so.

The outstretched plain between the mountains and the sea, is not perfectly level; it slopes very gently seaward, and consists of a great depth of diluvial and transported material, all small where visible. At Pæstum, and for a considerable distance round it, the fawn-coloured aqueous tufa, of calcareous matter filled with the impressions of recent plants of a paludal character — great arundos, alder leaves and twigs, &c.—is found horizontally, everywhere at from 6 to 12 feet beneath the surface, and no doubt overlies the limestone that supports the whole plain.

Of this tufa, the majestic, solitary, and awe-inspiring Doric temples were built, with the town walls of huge ashlar all laid dry, that alone remain of what was once a populous city. Upon the dreary winter afternoon, on which I examined its ruins, no sign of life enlivened the desolate plain, but a flock of screaming green plover; no sound was heard but the wind that sighed through the sedges and the distant and dismal howl of some goatherd's dog.

The formation of this tufa seems to indicate the upheaval of the great plain at a recent geological period. The lime has no doubt come, from the ground-up, and dissolved limestone, of the cretaceous formations, that constitute the lower mountain range, between the great mountain masses, of Apennine limestone and the sea. The latter occasionally comes forward in grand developments as at the Tusciano (Fig. 123), where the ranges under Monte Polveracchio show beds of vast extent and thickness, with a nearly horizontal strike parallel with the coast, and a noble sweeping curve, dipping steeply inland (about 30°) or towards the N.N.E. Far beyond are high and jagged peaks, and a lofty sierra, thinly covered with a hoary head of snow, bounds the horizon, and glitters against the cold grey sky. Some twelve miles further south, the Salaris crosses the plain in a deep channel with heavy slob banks in the diluvium, about 270 feet wide, and with a rapid current, and turbid mud-stained water, of about 20 feet in depth. It drains a large area, in a course of more than a hundred miles, and the quantity of calcareous matter, both in solution by carbonic acid, and in suspension as mud, that it constantly brings into the sea, must be even now producing very sensible effects upon the coast, *the dissolved lime forming the cement*, that rapidly agglomerates and hardens the calcareous mud into stone. This process seems also to be that, upon which the formation of the calcareous breccias found in such vast masses in the Apennines, has depended.

Looking eastward towards, the valley through which the

Salaris debouches upon the plain, between Eboli on the north, and the Bosco di Persano, the synclinal beds of limestone are developed upon the grandest scale (Fig. 124): from a to b is probably not less than twenty miles, and the same beds can be traced by the telescope at either side. Above and between these the upper limestone, of the collines, seems to lie unconformably. Upon one of the most prominent of these towards the south are perched, Capaccio Nuovo and Capaccio Vetico, with the great adjacent monastery; in all which, I was informed, the earthquake had been severely felt, but no considerable damage done. The general aspect of these branches of the Apennines, as I look back and into their recesses, is one of extreme confusion and dislocation, produced by long-continued and reiterated elevatory action of the most violent character, of which no just idea is given, by the surface configuration even of the largest maps, such as those of Bachler D'Albe and Zannoni; and such as it would require years of labour from the field geologist to analyse and describe.

At Pæstum I examined the ruins of the temples with care, for evidences of the shock, but they presented not the smallest indication of dislocation. Formed, as they are, of extremely massive blocks, laid without cement, and with all the top weight due to Greek Doric architecture, few buildings could be by form and structure more amenable to "promptings from beneath." A careful examination, however, led me to conclude that since their foundation, they had suffered nothing by earthquake, not even to the opening of a joint — a sufficient disproof of the common tradition, that Pæstum was deserted and reduced to ruin, by reason of the earthquakes that desolated the plain.

So far from the truth is this, that Capaccio, up in the mountain to which the Pæstians are said to have migrated from the plain, has been repeatedly dislocated, and, apparently, the ruin of the whole town produced the founding of Capaccio Nuovo; the other and older being now nearly without inhabitants.

In fact, from whatever centre earthquake movement originates, along the mountain axis from Calabria northwards towards Melfi, &c., its spread is greatest and most rapid, in the lower and denser limestone of the higher central chain, and here, at the western seaboard plain, is almost limited by the line of outlying cretaceous collines; the blow transmitted from which through ten or fifteen miles of soft porous tufa and loose material, principally deep calcareous clays, is completely buffed and lost, before it reaches Pæstum and the shore.

The family and servants of the landowner at Pæstum, a great number of whom I found collected at the Casone, were unanimous that the shock of 16th December was simply "oscillatorio," and in direction "levante ovvero ponente," that they had felt but one shock, and had heard no noise. On causing some of them to point out separately for me, from the balcony by the hand, the direction in which they deemed the shock traversed, and comparing it with the azimuth compass, I found it was very nearly E. and W.

They said the dogs (of which they keep a great number of large formidable animals to take care of the buffaloes, &c.), had barked violently and universally, for a good while before they felt the shock. Most of the people were in bed. They could give no information as to the time beyond crude guesses. The shock had been felt, they said, far worse at Agropoli and below it, than with them; naturally so, for the limestone mountains jut out into bold promontories and come down to the sea at or near that town. At Capaccio Nuovo they said the shock had also been felt from E. to W., and more from the north, than with them at Pæstum.

CHAPTER V.

ENTRANCE TO THE MOUNTAINS AND TO THE REGION OF RUIN—EBOLI—CASTELLUCCIO.

THE buildings of the Locanda di Vozzi, at Eboli, are those of an old, suppressed monastery of the Riformati, of great size, nearly square, and not far from cardinal, but two stories above ground, and extremely well circumstanced for observation.

The front with the campanile (Photog. No. 124, *bis*) along the main road bears 23° W. of N. Its internal construction, generally, consists (block plan, Fig. 127) of a central interior court, surrounded by a double range of rooms with a corridor between, running right round the whole building. Two of the corridors, and a few of the largest apartments, are vaulted with brick or rubble arching; the others, as well as nearly all the rooms, are timber floored. The rooms are chiefly small, having been formerly monks' cells, and separated chiefly by walls of one brick thick. The external walls generally are 22 inches thick, of stone. There are four huge external buttresses on the south side, up to the level of the first floor, built after the shock of 1851, which shook the building a good deal. There is a stubbed old cylindrical tower at the N. E. quoin, and a small external terrace and building at the S. E. one. The

WEST FLANK, EBOLI.

EAST FLANK, EBOLI.

N. and S. corridors are vaulted, the E. and W. ones timber floored.

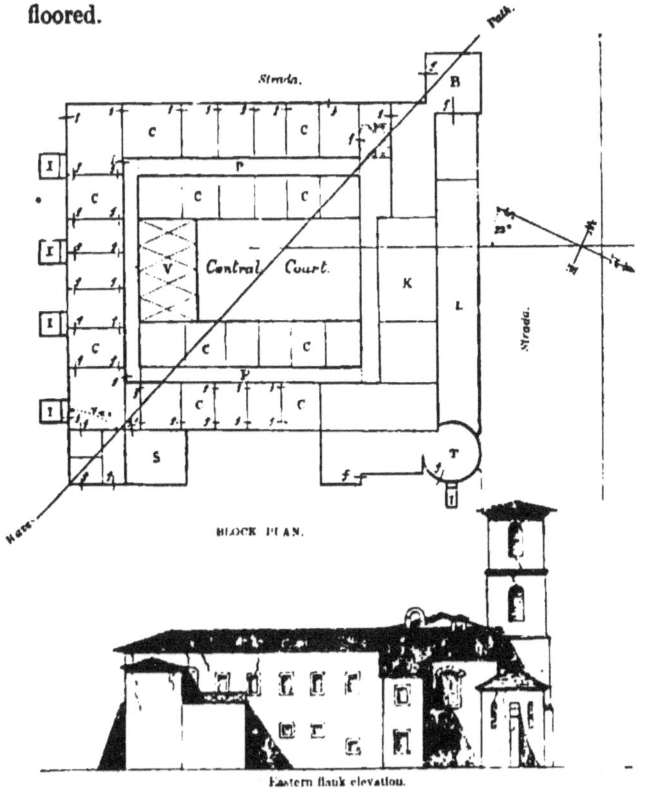

Fig. 127.
SKETCH OF LOCANDA DI VOZZI AT EBOLI
(*formerly a Monastery of Riformati*).

B, Campanile. C, C, Cells or chambers. I, I, Stone buttresses. K, Kitchen—common rooms, padrone, &c. L, Fore court. P, Vaulted corridors. V, Vaulted chambers. T, Tower. S, Small open terrace. *f* | thus marked are wall fissures.

The whole building is fissured in almost every part, and yet not uninhabitable, all the fissures being narrow, and but moderately inclined, and its peculiar cancellated structure has, by the help of the mid lines of corridors, kept the roofing uninjured nearly.

The fissures that are most instructive, are those in the external quoins, those on the soffits of the arching, and at the junction of the cross walls between the rooms (or cells) and the external walls.

The fissures resulting from the shock of December last are readily distinguishable from those of 1851. The walls have been all limewashed more than once, in the interval of time between, so that the former fissures are filled and obscured, those of December last, new, clean, and empty. The largest fissures are open 0·3 to 0·5 inch in 10 feet. They are wider, larger, and more numerous on the S. and E. wings. The width of similarly circumstanced fissures on the E. and S. wings are on the average of eleven pairs as 7 : 6·5. The angle (θ) made by the path of the wave with the east flank wall, therefore, was

$$\operatorname{Tan} \theta = \frac{6 \cdot 5}{7} = \cdot 93 \text{ nearly,}$$

or $\theta = 43°$, but the building is ordinal 23° W. of N., adding this, we have the azimuth of the wave-path at Eboli $= 43° + 23° = 66°$ W. of N.

None of the photographs (Figs. 124, 125, 126, Coll. Roy. Soc.), were taken sufficiently near, unfortunately, to show with distinctness any external fissures in this building, nor to show the S. W. angle at all. Referring to the block plan and elevation (Fig. 127), however, some large fissures in the external walls of the projecting buildings over the terrace, and, generally, in that portion of the structure, give an angle of slope with the vertical $= 15°$ to $16°$, and allowing for the change due to the ordinal of the building, indicate an angle of emergence for the wave-path $e = 18°$ from the E.

The church was the only other building I could find at

Eboli giving distinct indications. It is almost cardinal, and shows some pretty extensive sloping fissures in the walls of the apse, behind the altar, and in the N. wall of the nave, these being respectively nearly orthogonal. The apse fissures are comparatively insignificant, and from the form of the walls difficult to compare with those of the nave. The general deduction as to wave-path, however, coincides pretty nearly with that from the Locanda, being E. to W. and some degrees to the N. of W. The slope of the fissures in the N. wall of the nave gives for the angle of emergence $e = 21° 30'$ from E.

The people generally at Eboli, and more particularly those of the family of the Padrone at the Locanda, state, that they experienced both shocks from E. to W., and that there was a considerable amount of vertical movement (sussultatorio).

A very intelligent man living at La Sala, but who was at Eboli on the night of the shock, whom I met at the Locanda (but whose name, I regret to find, I did not note), agreed with their notions, but thought he had also felt a sharp jerking shock from the N. or E. of N. directly after both the first and second principal shocks. If this be so, it would probably be accounted for, as a reflected wave from the high hills that overhang Eboli to the N. and N.E. No one heard any noise attending the earthquake.

The level of the ground on which the Locanda stands, as given by reduced barometer observation, is 326·8 feet above the sea. Barom. reads 29° 72′, thermo. 50° Fah. at $9^h 25^m$, Nap. m. t. (26th February).

At the Locanda of Eboli, I had the good luck to fall in with Signor Palmieri, the engineer of the corps of Ponti e Strade, to whom I had a letter from the Intendente

Ajossa, and during the evening spent with him I obtained a great deal of useful local and other information, as to matters of fact from him.

He gave me circumstances I deemed conclusive, that the wave-path had been from N. to S. at Lago Negro and at Sapri, and at Laurino from the S.W. to N. E. I fell in again with Signor Palmieri near Polla, whence he was returning, and was indebted to him for first calling my attention to the instructive facts developed at the Palazzo Palmieri there, which he had just been examining.

About four miles from Eboli I crossed the Salaris by a grand old irregular bridge of one very large semicircular arch and several minor ones, and here first observed masses of the limestone pebble breccia, of the Apennine formation (the bridge is built of it), which, from this eastward, appears to overlie the limestone and underlie the deep diluvial clays and gravels of the valley bottom. The pebbles hereabouts are usually from two to four inches diameter, much water-worn and rounded, of a brown-grey and fawn-coloured argillo-calcareous rock, with a good many occasional pebbles of a tea-green, cherty, metamorphic slate, so hard as almost to resemble Jade. The cementing material is calcareous, and the interstices of the pebbles are filled with fine gravel and calcareous sand. It is a coarse but good building stone, and indurates much on exposure. *In situ*, this breccia here is stratified in great coarsely-defined and irregular beds, which generally approach the level in strike, though much tilted transversely to the line of valley.

At the fork of a small stream, the Merdarolo (or Pagliardo according to some of the peasants), where it joins the Salaris on the left bank, beds of limestone tilted almost vertically appear with a lithological character almost identical with our English lias limestone. They are unconformable with the enormous pile of limestone beds, nearly horizontal in strike along the valley, but dipping sharply to the south, which form the huge, shattered, and decussate precipices, rising to the summits of La Scorza or Monte Alburno. These summits shut in the Piano of Savannola, a singular irregularly oval-shaped, and mountainous table-land, of more than twenty square miles in surface, and, beyond it, I get occasional glimpses of still higher ridges and peaks. The range of Alburno must rise to at least 3,000 feet above the plain between Eboli and the sea, and is now (12th February) covered with snow, for about half the depth down from the top, wherever it can lodge upon its abrupt and precipitous flank.

For about the lowermost third in height on both sides, the valley is covered with smoothed, rounded, and sloping masses of diluvial clays and gravels, with huge angular blocks and boulders dislodged from above scattered here and there.

Noble natural oak forest, clothes much of this down nearly to the valley bottom, as in the days when Virgil wrote his third Georgic:—

> "Asper acerba sonans; quo tota exterrita sylvis
> Diffugiunt armenta, furit mugitibus æther
> Concussus sylvæoque et sicca ripa Tanagri."

Through these and below them, the rain has cut into these clays in a surprising manner, and in many places they seem as if subsiding bodily, from off the steep sides of the hills, and melting into the muddy flood of the Salaris, which, a few miles further eastward, unites its current with that of the Rio Negro or Tanagro, falling in upon its left bank, between the towns of Contursi on the north and Postiglione on the south, right under which are the ancient and the new post-houses of La Duchessa.

Here the first walls actually prostrated by the earthquake become visible (going eastward). The old posthouse is now a roofless ruin, of walls standing two stories (without floors) or about 30 feet in height, about 250 feet long by 45 feet wide, built of rubble limestone, with ashlar quoins. None of the stones of the walls exceed about 6 ins. × 12 × 12, or 18 at most; poor masonry, but not ill suited to seismometry.

The place was damaged by the earthquake of 1783, and rendered uninhabitable by that of 1851. Its general length bears 135° E. of N. The fractures and fissures produced in December are clear and distinct from the old ones. In the sketch (Fig. 128), the portions coloured black indicate the

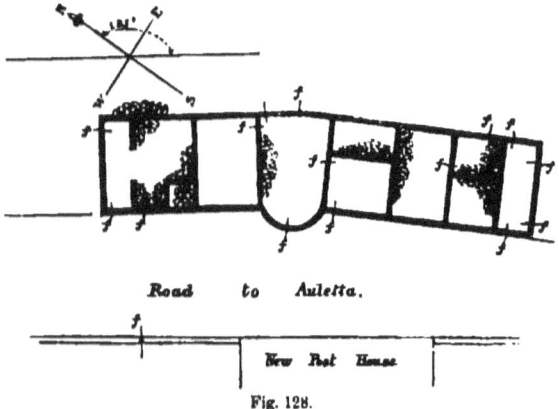

Fig. 128.

walls partly or wholly fallen; and the places of the principal measurable fissures are marked ff, &c. The fissures of the S. E. end when reduced, give 76° W. of N. for the wave-path, those at the opposite end 73° W. of N. the mean = 74° 30' W. of N. Those at the former end are pretty evenly and uniformly sloped, and at an angle with the vertical = 24°, giving that for the angle of emergence of the wave-path. Some of the fissures, however, gave an angle of emergence as steep as 32° or 33°.

This indication puzzled me much. The N. and S. direction of wave-path at Naples, and on the south side of the bay, led me to expect that I should find the focus of the shock had been somewhere at sea, under the Gulf of Salerno, or that of Polycastro; and this was supported by continuing to find that the wave-path from Salerno and southwards was W. and E. more or less. I was not shaken from this hypothesis, by finding that no traces of the incoming of a roller or great sea wave had been observed anywhere along the coast from Salerno downwards; but the finding these fissures at Duchessa, leaning off towards the south and east, seemed irreconcileable with any possible position of focus that could account for the observed wave-path at Naples, inasmuch as it could alone be the result of a focus to the eastward of Duchessa, and therefore far away from intersecting any line drawn nearly north and south through Naples. I summoned my faith in induction to my aid, however, and trusted to the witness of further facts to solve the mystery.

The postmaster here, had heard a sound along with the shock: he thought it came rather before the actual oscillation, which was undulatory also. He was standing up, in the room in which he slept, and described the sound at or preceding the first shock, " Not loud but a quick sort of deep hoarse buzz "—" Roco ronzio profondo vivace, ma piccolo." He could not tell whether there was any sound with the second shock or not—" he thought there probably was—but they were all in great alarm." One or two considerable fissures had been produced in the new posthouse. I took observation of the sun here to determine the magnetic declination, and give the result as worked out, but am satisfied I must have committed some gross error in the observation or note of it.

Duchessa, Lat. 40° 34' N., Long. 15° 11' E. (Feb. 13)
Hour angle at time of observation = 24° 6' 48"·15.
Sun's azimuth computed . . = 27° 42' west.
Sun's bearing by compass . . = 20° 0' east. (?)
_____
Magnetic declination . . = 47° 42' east.

There is no ground for assuming any serious local disturbance of declination *here*.

From a little beyond Duchessa, off to the eastward, and rather to the south, Sisignano is seen perched upon a spur of Monte Alburno, that runs in a S. E. and N.W. direction diagonally across the main valley and down to the Tanagro. The road continues to rise rapidly, ascending over the shoulder of this, at the little village of Lupino, and to the south or right hand. The vast pile of limestone beds of the Alburno is seen stretching away with nearly horizontal strike, parallel to the general line of the great valley, and dipping sharply to the west and south-west, at probably 45°. The transverse ridge is too much covered to enable me to prove, what I conjecture from its outlines, that it consists mainly of the breccia limestone, unconformably laid on the beds of Alburno.

After about six miles, the highest point of the transverse ridge is reached, near Lupino, and I determined its height by barometer, which reads, at 1·32 Nap. mean time, 28·78 inches. Thermo. 52° Fahr. (13th Feb.). This reduced, gives 1441·3 for the altitude. (See Appendix for particulars of all barom. measurements and reductions, &c.) The transverse ridge here, is a complete dividing barrier, upon the south side of the Tanagro, between the great valley of the united streams of the Salaris and Tanagro, and that of the latter river, into which I begin now to descend again rapidly, to within perhaps 400 feet above the bed of the river.

At Lupino, almost no damage was done. The few houses are low, well built, and not very old. The postmaster here was rather uncommunicative. "They had been severely shaken and much alarmed, but knew of no damage done at Lupino. I should find plenty six or seven miles further eastward." I can see with the telescope the old château on the highest part of Sisignano, overthrown and in ruins, and a good deal of damage in the place itself. Gualdo, with Terra Nuova, are above me on the south, but neither have suffered very much. About two miles further on I pass the Taberna of Urma, a small post-house, with a new and yet unroofed Capella close to it, which had just been built, and the mortar of its limestone rubble yet fresh and soft. It was a building of one story, about 30 feet E. and W. by 24 feet N. and S., and the walls about 17 feet high. These are fissured, at three out of the four quoins in such a manner, that the ends tend to come out. The fissures are widest at the east end. The axial line is exactly cardinal by my prismatic compass, and the fissures give a wave-path of 81° 30' west of north. The stones are large in proportion to the size of the building, and I can get no indication reliably as to emergence.

Numbers of fissured buildings now begin to present themselves, all indicating as I pass them a general east and west direction of wave-path. A little further on, after passing a torrent that falls into the Tanagro from the south, I look up an extremely steep lateral valley—Il Vallone Petroso—in which many loose surface blocks of limestone, show themselves to have been shaken from their positions and rolled over: from the road at this point there cannot be less than 3,500 feet vertical, of calcareous beds above me. The geological evidences of violent dislocation and elevation at all sides in the mountain formation are strikingly grand.

At the 58th milestone from Naples on the military road I am close under Castelluccio, a strange, immured, and gloomy-looking mediæval town, perched on the very crest of a solid, rounded, lumpy mass of limestone, showing little or no signs of distinct bedding, and with its sides so steep, that trains of loose stones lie in huge furrows, straight up and down its flanks here and there. The Tanagro flows at the opposite side or round to the north of this hill, while its tributary from the Vallone Petroso winds round the foot of the enormous rock, (see Photog. No. 13, Part I.,) to join the former to the N. E. The town and its eminence thus stand upon a sort of peninsula, rising more gradually from the main valley upon the westward, and having the longer axis of the rocky mass nearly in an E. and W. direction.

Although close enough to see the joints of the masonry in its walls, through the keen clear air, with the naked eye, I found it would require four hours' time to climb up to the town; and learning that it had sustained but very little damage, I did not attempt to lose time in the ascent, but scanned it narrowly with the telescope in a fine light. Not a single fissure was visible in the N. W. or E. sides of its external walls, which, although they look like those of a large fortified mediæval town, are in reality only the rear walls of the houses turned outwards, and built closely together, and upon the very edge of the steepest escapements of the rock. Such are the characteristics of many of these most interesting old towns.

But I can see that the top of the cupola of the highest campanile or tower in the place, probably that of the church, has been broken off short and is gone, obviously by the direction of the fracture, by a force from the eastward and a little south, or about 70° W. of N. by estimation (c, Fig. 129), being, as figured, at the east end of the cupola, which was an hexagonal, Saracenic sort of small dome, of limestone, obviously modern.

As I move round the base of the rock, I can see thus much reason for its comparative security, that the mass of solid limestone upon which it rests, presents its long way to the length of the valley, and lies nearly E. and W.,

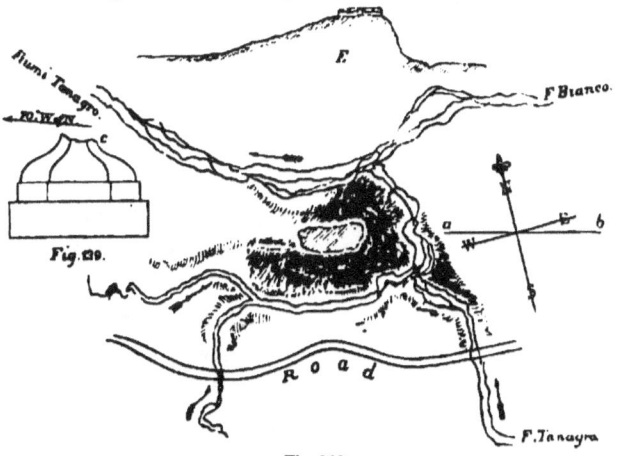

Fig. 130.

with its steep end towards the east, and buttressed away to the westward by a longer slope, as in the section E (Fig. 130) taken in a line with *a b*.

Quite beneath Castelluccio, between the road and the torrent, that rushes to the Tanagro round its base, and some 100 feet below the road level, is a solitary church, La Chiesa d'Incoronata. It stands upon a low lying spur of deep diluvial clay and gravel, upon nearly the edge, that scarps sharply down, covered with natural oak and hazel, to the torrent of the Petrosa, and is greatly damaged; the whole of the east end having fallen out, carrying much of the roof with it. Upon descending below it, through the woods, I find that the deep diluvium above rests upon argillaceous beds, which are nearly vertical, and strike across the valley in a N. W. and S. E. direction, and so are almost parallel with the ridge of Sisignano and Lupino, already passed, and which appear to be wholly unconformable to the limestone breccia of Monte Carpineto; the subordinate mountain, to the continuation of the N. scarp of Monte Alburno, and which lies S. and S. E. of Castelluccio.

This church is a poor building, the walls about 15 feet high and $2\frac{1}{4}$ feet in thickness, of coarse limestone rubble, covered with a heavy tiled roof upon gross, ill-framed timber. The north wall had, in part, long leaned outwards (as I was informed by the priest), and a portion had fallen towards the north; but all the remainder of the east end had fallen outwards, or in a general direction of the line a to b (Fig. 131), and a much larger portion of the roof, as indicated by the irregular line $c\ c$, had come down and fallen within the walls. In both the north and south walls were some fissures $f f f$, which, together with the general direction in which the mass of dislodged material had been thrown, indicated a wave-path of from 80° 30′ W. of N. to nearly due W. and E. The mode in which the roof had come down, seemed to indicate a considerable amount of emergence in the wave-path, from the eastward, but

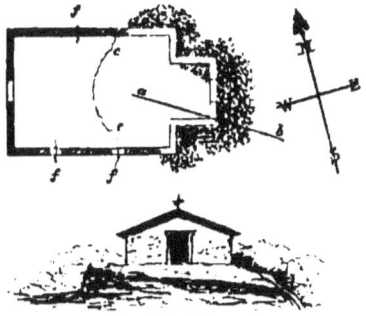

Fig. 131.

how much, neither it, nor the direction of the fissures, in the coarse rubble, and through the heads of the windows, would indicate.

Monte Carpineto is part of a subordinate range, that stretches in a N. E. and S. W. direction, as far as Salvitella, northwards, almost at right angles to the great north scarp of Monte Alburno; the Tanagro piercing through it nearly at right angles, about half way between Petina, on the south, and Salvitella; and the much larger stream which comes away from far north of Muro and Bella, in the lofty recesses of Monte Croce, and joins into the Tanagro at Castelluccio, and properly should be called Tanagro, taking the name of the Bianco. The descent is extremely rapid from about the 59th milestone, or half a mile beyond; and as the new valley begins to open, I catch the first glimpses of Auletta, and even at this distance perceive the terrible evidences of the overthrow it has sustained.

Monte Carpineto once passed, the general direction of the main valley changes, and I descend into a deep and nearly closed-in hollow, between smaller lateral valleys running nearly north and south, and opening in common upon the course of the Tanagro, in which Auletta and Pertosa, each perched upon a separate spur or colline, and each in the mouth of a separate but closely adjacent valley, are situated.

Upon the south-east, this hollow is completely barred and closed in, by the mountain of Taliata at the west, and the ridges of Monte Sarconi on the east, which abut upon each other, with nothing intervening but the tremendous cleft, through which one portion of the Tanagro forces its way from the north end of the Piano di Diano; while another portion of it, disappearing there, finds its way by a subterraneous channel, nearly parallel with that *sub dio*, and both meet again at Pertosa; the subterraneous waters discharging from the mouth of St. Michael's Cave, at a level of about 200 feet above the open bed of the river, opposite Pertosa, and turning the wheels of some old Catalan iron forges, before falling into it.

Although from this point the Tanagro pursues the same direction, as lower down, nearly from E. S. E. to W. N. W., the whole character of the valley itself has changed since passing Castelluccio, it is no longer a great E. and W. valley, but an irregular deep hollow, produced by the abutting and inosculation, of several secondary mountain ranges and valleys, running more or less N. and S., and gradually shutting up the hollow, until Auletta and Pertosa seem enclosed within it.

CHAPTER VI.

ENTRANCE WITHIN THE MEIZOSEISMAL AREA—AULETTA —GREAT EARTH FISSURES.

I HAD no sooner passed into this hollow, than it became evident, that all at once, I had got within the radius of formidable earthquake violence: on every side ruined and prostrate buildings presented themselves. Descending towards Auletta, I can see Buccino greatly elevated, and some six miles to the north: many of the people here call it Bugille, pronounced like French, without the final vowel. This corrupt pronunciation of names is frequent, and renders recognition by maps often difficult; the same name is often pronounced half a dozen different ways by as many persons—an existing example of that jargon of living tongues, betwixt closely adjacent places, of which Sismondi gives so vivid a picture in those very regions, in his 'History of the Literature of the South of Europe in the Middle Ages.'

At half a mile in a right line from Auletta, I pass a ruined house, of one story, of about twelve feet in height, and about twenty feet square in clear of walls, which are two feet thick, pretty well built, and not above ten or fifteen years of age. This and a much larger building, a sort of farmstead, at the opposite side of the road, with a great spread of gable, and front parallel to this little house, present evidences of great violence here, and extremely steep emergence of the wave. The house (Fig. 132) is fissured from wall plate and ground, in the north

Fig. 132.

and south walls, near the quoins, and over the entrance door, the lintel of which, a single slab of limestone of about 12 in. by 22 in. wide, is fractured right through by the rocking of the wall at either side of it. The roof is all fallen in, except one purlin, which is still *in situ*, and proves, on examining its ends in the gable walls, that the east gable, and the roof along with it, all came to the eastward, at the first movement of shock, drawing the purlins from their sockets in the west gable; some were drawn quite out, and the roof at once fell in; others returned by the back stroke of the wave, and drove the ends of the remaining purlins, back again into their sockets like battering rams, throwing out portions and dislocating the west gable. The north-west quoin has had a long wedge-shaped mass projected right outwards, showing a very steep angle of emergence. The dislocations generally, here and at the opposite side of the road, give evidence of a wave-path from E. to W., and an angle of emergence of upwards of 45° with the horizon. I did not measure any

AULETTA, SHOWING THE DIRECTIONS OF THE LANDSLIP, AND LONG FISSURES IN THE SOIL.
EYE SKETCH

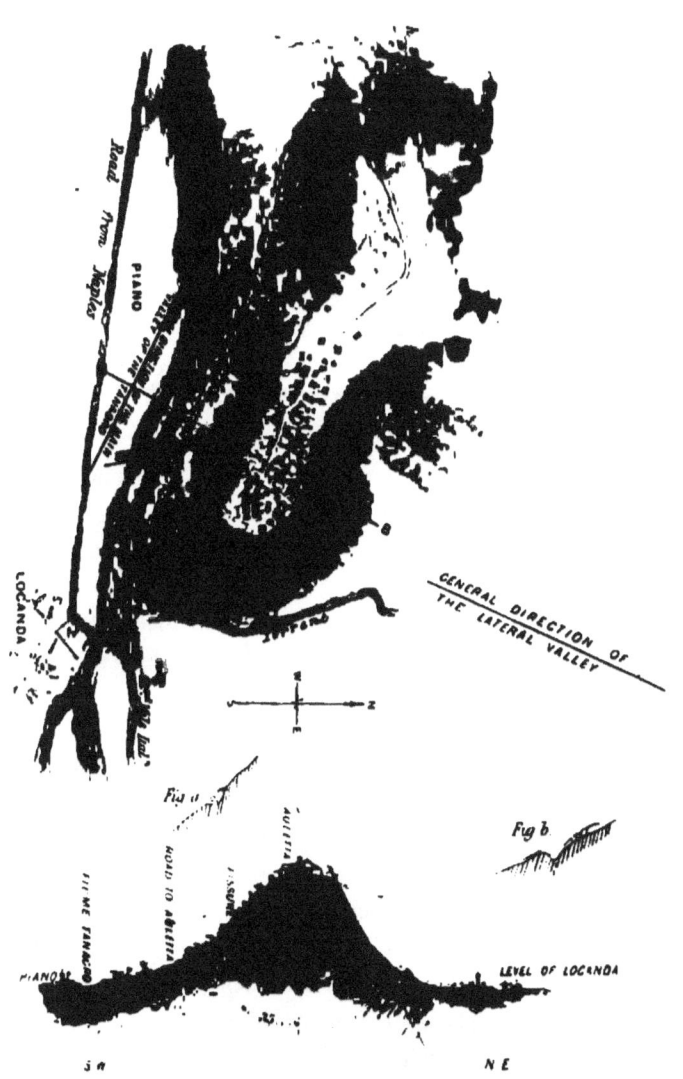

TRANSVERSE SECTION A TO B.

angles, however. In this house a whole family were crushed beneath their humble roof: they had been asleep, and, awakened by the first movement or by the noise, had tried to escape, but the broken lintel, had jammed the obdurate oaken door, just within which their bodies were found collected, beneath the mass of tiles and timber.

Auletta stood upon an elevated knoll, jutting with a S. E. direction from the E. slope of the mountain, of its own N. and S. valley, with the Tanagro sweeping past its base to the southward, and joined by a small torrent on the right bank from the Auletta valley. The bottom of the proper valley of the Tanagro here presents a broad level plateau of a mile or so across, upon which, at the river and beneath the base of the knoll, are some other houses, &c., with a poor and now half-ruined locanda, that also go by the name of Auletta. The town itself on the top, had about three thousand inhabitants, some fine large houses inhabited by official persons, and a mass of poorly built ones, all of the nobbly limestone rubble. It is mediæval, and was fortified in the middle of the sixteenth century, the castello alone now being the only visible fragment of this.

Referring to the eye-sketch (Fig. 132), the spur or elongated knoll upon which it stands has a general direction of N. 60° W. The N. E. side is extremely steep. From the highest point of the remains of the castello I found by the theodolite the general angle of the scarp to be 41° from the vertical. The opposite, or S.W. side, slopes much more gently. The N.W. junction with the mountain range, is a little depressed below the summit at the town, and undulatory. The spur consists of coarse calcareous breccia, in heavy irregular beds, with a nearly N. and S.

strike, and dipping 35° to the S. and S.W. These are visible here and there all over the steep scarp, and at several points in ascending from the locanda, along the road at the S.W. flank. The character of the rock, which is the average of most of the breccia hereabouts, is seen in Photog. No. 133 (Coll. Roy. Soc.), taken from within forty or fifty feet of the rock; the pebbles are from three to ten inches diameter, and extremely round. The steep scarp is almost bare rock, but that to the S.W. is covered to within one-third of the height, or less, from the top, with diluvial clay, and olives grow over the whole slope. The clay is perhaps thirty to fifty feet in thickness near the base, between the road up towards the town and the Tanagro, thinning off to nothing as we ascend.

At the bed of the Tanagro, below the bridge that passes over the great military road, barom. reads 29·76 in., thermo. 51° (13th February). At the locanda, which is on the level of the valley piano or bottom, (14th February,) barom. 29·50 in., thermo. 52°; and at the summit of the town on top (13th February), barom. 29·41 in., thermo. 48°. These reduced (see table, Appendix), give for the respective levels above the sea—

        Bed of the Tanagro . . . . 576·0 feet.
        Valley piano, or bottom . . . . . 647·6 „
        Summit of Auletta . . . . 889·5 „

The spur is therefore 242 feet nearly above the valley bottom. It can be scarcely half a mile in a right line across the base in a direction A to B of section, and perhaps two miles the long way, in its projection from the main slopes.

A knowledge of the magnetic declination, as affecting all my determinations, rendered frequent observations to ascertain its amount important; unfortunately, from the season of the year, and inclemency of the weather in these mountain regions, the observation of the sun or pole star was practicable but seldom. I therefore, in addition to such solar or stellar observations as were possible, took magnetic bearings, from many elevated known points, of others visible from them, and recognizable again upon the two great maps of the country, Zannoni's and Bachler D'Albe's, so that, by comparing the observed bearings with those of the maps, the declination might be checked. I took the following bearings at Auletta:—

Villa Carusso bears from Auletta 7°   W. of N.
Contursi . . . . 43° 30' W. of N.
Castelluccio . . . . 56° 3' W. of N.
Petina . . . . 93° 0' W. of N.
Caggiano . . . . 89° 0' E. of N.
Pertosa . . . . 82°   E. of N.

These give from 14° to 14°·50' declination W.

On the 13th February I was able to see the sun's disc, though not perfectly clear, and take observations. At 9h. 0m. Greenwich time by chronometer, the centre of the sun's disc bears 24° E. by compass. Taking Auletta to be in lat. 40° 30' N., long. 15° 23' E.

The hour angle at time of observation = 32° 25' 34"·50
The sun's azimuth . . . . = 36° 20' East
Ditto by compass . . . . = 24° 0' East
─────────
Declination . . . . . . = 12° 20' West

This result may be in error a degree or two, for the latitude and longitude are had from map measurement, and the cloudy mist that hangs at this season all the forenoon, over the chill waters of the rivers in these mountain valleys, precludes good observation. The declination here is not, therefore, far from the same as at Naples.

Auletta has suffered much from the shock, most so, at its highest portions, and upon the N. E. side of the summit, where the most substantial buildings stood, and upon the S.W., where were some of the poorest and worst.

The general appearance of the locality is seen in Photog. No. 134 (Coll. Roy. Soc.) from the road sloping up to the town, and Photogs. Nos. 135 and 136 give two views nearly at right angles to each other, of some of the most instructive buildings at the upper part of the town (No. 135), looking, about N.W.

The propped wall, in both Photogs. is the same, and is square to the four parallel walls, which in No. 136 are observed all shorn off and thrown, from the western quoins. This is a good example of the parallelism of angle at which such fractures form in *large* masses.

At the lower part of the town, and upon the slope towards the N.W. leading to it, the fractures all indicate a very steep emergence from the eastward—upwards of 45°; but upon the summit, the buildings show a lower apparent angle of emergence, and *greater dislocation*. This obviously arises from the fact that, as the wave-path hereabouts was in some direction from E. to W., and therefore diagonally transverse, to the narrow or thin direction of the spur, upon which the town stands; the mass of *the spur itself vibrated*

AULETTA.

AT AULETTA.

with the blow as an elastic pendulum, so that its proper motion at top, which was quam prox. horizontal, was added to that of the emergent wave, thus reducing the direction to one of less emergence, and increasing the range and velocity of movement upon top. The great majority of the fractures at Auletta, indicate a wave-path E. and W. by compass; those of the castello, which is prostrated to the level of the first story, and stands upon the very brink of the precipitous side, when reduced give one of 115° W. of N.; but the buildings are not isolated, and the direction has been probably perturbed by this, and by longitudinal vibration in the mass of the spur on which it stands. On the whole, the wave-path here is E. and W.

The effects, of the form, elevation above its base, substance, and direction with respect to wave-path, of the collines or spurs, upon which so many of these towns are perched, in modifying the results of the shock upon them, are strikingly seen here, us well as at Castelluccio, which we have passed, and at the little village of Petina, which, from a point between this and Pertosa, I can descry with the telescope, perched high up, upon the south scarp of Monte Alburno, at least a thousand feet above me. It stands upon a level sort of short, stumpy, buttressed spur, jutting out from the steep mountain slope, which in form is like a piece of artificial earthwork, the little town standing upon the level platform on top, with a steep scarp in front of it, the mountain rising abruptly behind, and the scarp sloping in and getting lost in the mountain sides to the E. and W. of the town. The terrace upon which it stands, however, is not earth, but solid limestone—a projection, of the great horizontal strike of the beds, of the great scarp of Alburno, as in Fig. 137, which dip pretty sharply to the south. I find upon inquiry here that Petina, which is just five

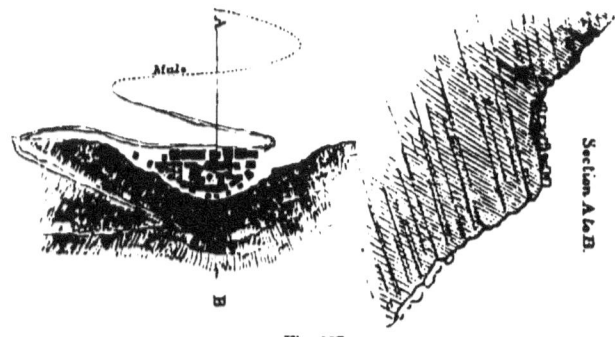

Fig. 137.

Italian miles S.W. of Auletta, and about six from Pertosa, has nevertheless suffered absolutely nothing, although these latter towns are in great part prostrated.

The immunity of Castelluccio from injury arose, as I before remarked, from the *long dimension* of its well-buttressed knoll being *opposed to the line of shock*, as well as to the barrier interposed between it and any shock coming from the eastward, by the mass of vertical breccia beds to the east of the town.

Petina has owed its immunity, first to the peculiarly strong form of the terrace upon which it is perched to resist any vibration in the mass itself, secondly, to the fact that from where I stand at Auletta Bridge, there is about 1,000 feet of piled-up limestone beds between me and Petina, so that any shock emergent at a steep angle, here or further eastward, must have passed up transversely through all these successive plates of variable hardness, and none in absolute contact with each other, and so the *vis vivâ* of the shock be enormously reduced before reaching the elevation of the village. This is probably also the reason why huge masses, of the towering and ruiniform crags, that form the highest summits of Monte Alburno, far above Petina, have not been brought also toppling down into the valley. As it is, however, I find, according to the information of the guard of gendarmes here stationed, that there have been some heavy falls of rock in the Vallone Petroso, and at other places along this end of the scarp of Alburno.

At Auletta it was alleged to me by these gendarmes, who arrived there within a day or two after the shock, and whose statements were confirmed, by those of some dozens of the poor inhabitants, that large and long fissures had opened in the earth around the town, but that they had since all become closed again, and they doubted that they could now be seen. I gave much scrutiny to this, and having got the corporal of the guard to come with me, he pointed out the place where he had observed one of the largest of these fissures, to the north-west of the town, amongst olive grounds, and in deep clays (at F, Fig. 132). After some time I found myself unmistakeable traces of the fissure, in a continuous sort of little narrow trench, about 12 or 15 inches wide, at the widest places on the surface, but generally not more than 8 inches wide, and of a blunt V shape in cross section, with rounded edges, and not above 8 or 9 inches deep, as in Fig. 132. The whole interior surface between the lips was free from vegetation, which grew, in many places, close up to the edges, and corresponded on opposite ones. There was no denying the evidence of a recent fissure, filled up still more recently, by the slow sinking together of the sides, and by the washing in of earth by rain. I was enabled to trace it along the surface, with but occasional breaks of continuity, where the rain had washed illuvium transversely and filled it, or where the ground had been tilled between the olives, for more than a quarter of a mile. I was informed by the same soldier (whose testimony I had thus proved trustworthy) that he had himself traced the fissure F for nearly two Italian miles in a west and south-west direction, which was one generally *coinciding with the horizontal contour along the slope of the hill side.* Returning back to whence we started, I found two divergent fissures (as figured), and traced one of these in a S.W. and S. direction for some hundreds of feet, down through the olive orchards parallel to the road up to Auletta. The one from h to k led me to try to follow it in a contour line, along the S.W. slope of the town, and without much difficulty I found it again, where the earth got deeper, and traced at several points, but not continuously (much matter having been washed across it on this steep slope), the fissure f. It was a similar little trench to the former, but as in Fig. (a, 132), *one side* of the V greatly *higher than the other*, smaller in size, and harder to trace than the preceding from the want of surface vegetation and the pebbles rolled into it. In all these it was manifest, that *the fissure was the evidence of a great earth slip*, and had *resulted, not from any direct rending asunder of the ground or rocks beneath it, but that the clay masses had when shaken violently upon the inclined beds of rock upon which they were superposed, slid down bodily by gravity, and parted off from each other at these fissures.*

The fissures, by their direction, perfectly sustain this view, but are absolutely opposed to the idea of fracture, either by shock or by unequal or local sudden elevation or depression of the subjacent formations. For the great *direction in length*, of the fissures, is *not far from that of the wavepath here*, while it is everywhere but little removed from one, *transverse to a line up and down the slope of the hills*.

The people generally stated, that these fissures were at first, from one to two palms wide, at the widest, and that in steep places, one of the lips was about a palm above the other, and that in one place, the depth had been probed with a rod to nearly thirty palms. The fissures evidently had in no case run down plumb into the soil, but sloped in the same direction, but with a less angle to the vertical than the hill side on which it was found.

There is hence nothing very surprising in their occurrence. They are all in deep soft diluvial clays, of great specific gravity, very fine grained, and almost free from gravel and stones —a tenacious clay loam, that when wetted gets at once into a sticky paste, and soon runs into a greasy cream, so that a very small declivity and moisture alone, produce frequent land slips. Thus on the road leading from Auletta, along the slope of the hill, towards Villa Carusso, (which is also that to Salvitella and Vietri,) on the N.W. side of the valley, I observed a place where a slip had occurred, upon a bed of not above 12° or 15° slope, which a year or two since had carried away the whole road (here on side cutting), and lowered its surface about 4 feet, for some 400 feet in length, carrying with it the telegraph poles, still standing fast in the soil.

But a small effort of vibratory movement, therefore, must be sufficient partially to dislodge masses of such material, on much steeper slopes. The average slope of the rock, beneath the soil of the great fissure here, cannot be less than 25° to the horizon, and may be much more. That of the fissure f is still steeper. The "work done" by the shock, in *originating* the actual transport of material, due to these fissures is not very large ; it amounts merely to reducing the friction and coherence of the mass, so as to permit the descent, through probably not more than 3 inches vertically, of several millions of tons of earth, the movement of descent being produced by gravity only.

The rain cuts these deep clays, here, (as everywhere in these provinces) into deep ravines or "nullahs." Some which I crossed in walking from the town of Auletta to Villa Carusso, which I next visited, were 30 to 40 feet in depth, with sides so steep and soil so greasy, that it was with great difficulty they could be crossed, and were it not for the hold on the brushwood at the sides, could not be ascended at all.

CHAPTER VII.

VILLA CARUSSO—FIRST DETERMINATIONS MADE OF WAVE-PATH'S EMERGENCE—PERTOSA—SOUNDS HEARD.

THE Villa Carusso is a large proprietor's house, visible from Auletta summit, at a distance of about 1½ mile in a right line; and seeing by the glass that it was a strong well-built house and nearly cardinal, I resolved to examine it. It is in general plan a parallelogram, with four towers at the angles, and a projecting sort of porch over the front and rear entrances.

Its longest axial line is 110° E. of N. The western end is probably 400 years old, part of an ancient semi-fortified château, and the towers here are of decayed and very inferior masonry. The remainder is modern, and of tolerably good uncoursed rubble limestone masonry, the stones not above 10 to 15 inches average bed, with dressed ashlar window and door jambs, and some Italian brickwork in the arcades over the south entrance. (See Figs. 139, 140, 141, and Photog. No. 137 *bis*, Part I.)

The walls at the west end appear to be about 2 feet 9 inches in thickness at about 10 feet from the bases, as I found by climbing up to one of the window apertures, that shown in Fig. 141, north elevation of the tower, V, Fig. 139.

No one was to be found, and the house was tenantless, so I could only make external examination.

The only fissures are those marked ff, &c., in the west end, all in the ancient towers, and two fractures in the slender brickwork of the southern arcades. The towers have been thrown westward, off from the body of the building, and the fissures at the reënterant quoins are $2\frac{1}{2}$ inches open at top, and run nearly from top to bottom of the walls, which are about 32 feet in height to the eaves, and the whole of the west end is out of plumb. The fissures in the south front are seen in Photog. No. 137. Their direction has been determined and limited at the quoin, by the form of the buttress built into it, for nearly

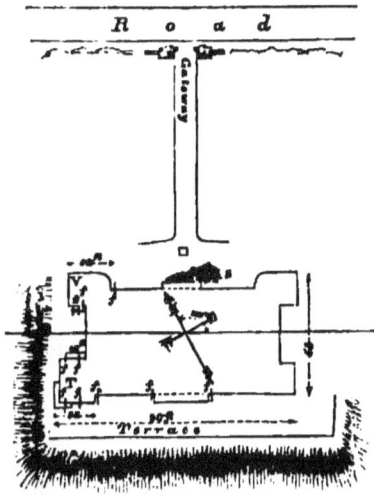

Fig. 139.

half its height, and projecting at the base, as seen in ground plan, Fig. 139.

The fissures at the internal flanks of the tower, that, T, (ground plan, Fig. 139), looking south, and that, V, looking north, are pretty exactly sketched in Figs. 140 and 141.

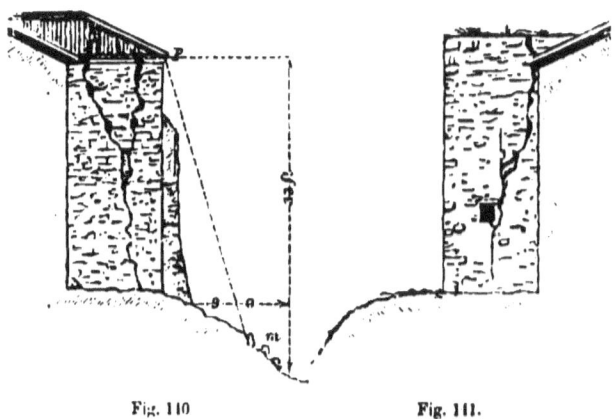

Fig. 140.   Fig. 141.

These fissures slope about 30° from bottom to top towards the east, but they, like the preceding, have been rendered more nearly plumb, by the action of the buttresses and other disturbing causes acting in the interior, which I am unable to examine particularly.

Some roof-tiles have been thrown from the eave at the west end, from p (Fig. 140), and are lying at m, from 7 to 9 feet horizontally from the base of the wall, having fallen vertically 33 feet.

Taking the horizontal range of projection here, $a = 9$ feet, and the vertical descent from the point p, $b = 33$ feet, and assuming the velocity of projection V to be 13 feet per second (which is about what it proved to be elsewhere in this neighbourhood), and solving for $e =$ the angle of elevation of the projectile which is here equal to the angle of emergence, we have,

$$\text{Tan } e = \frac{2 \Pi \pm \sqrt{4 \Pi (\Pi + b) - a^2}}{a}$$

in which Π is the height due to V, and finally obtain $e = 68° 25'$, for the angle of emergence, or that of the wave-path with the horizon; but as the tiles were fastened more or less, some force was expended in detaching them, so that the angle e, was rather less than this, and may be estimated about 60° or 64°.

The house was approached by an arched gateway, of rubble stone-work plastered, as seen in Photog. No. 138, looking south from the military road. This arched structure is heavily fissured, right through the crown of the arch, showing most at the south side, and in three diagonal fissures on the north side of the structure. The largest fissure is about 4 inches wide at top. The gateway is 12 feet wide and the height to the soffit is 16 feet; each pier is equivalent to about $4\frac{1}{2}$ feet square in horizontal section, and buttressed by the fragments of walls proceeding from them at each side. These fissures, like those of the main building, indicate a wave-path not far from E. and W., and a direction from the eastward—on the whole, about 100° west of north. The gateway fissures have had their directions mainly determined by those of the voussoirs of the arch, which is of brick; by the relative support given by the side walls as buttresses; and by the *rocking* of the piers as the shock passed through them; so that no inference can be drawn from them as to the angle of emergence.

In the Photog. No. 138 the rere or north front of the villa is seen, and above the projection at the central part

THE PORTE COCHERE, ON THE MILITARY ROAD, VILLA CARUSSO, NEAR AULETTA.

(which corresponds to that above the "Portone" of the south front) stood, before the earthquake, a wooden sort of verandah, framed, as shown in Fig. 142, and covered by a

Fig. 142.

prolongation of the slope of tiling of the roof. The whole of this had been thrown down, and was lying on the ground at and about the point *s* (in Figs. 139, 142, and Photog. No. 138) in dislocated fragments, from which, collecting and laying together the timber-work, I was able to reconstruct the design of the fabric.

The shock emerging from the eastward had swept over the vertical posts towards the direction of the dotted lines *n n n n*, and the tiling of the roof had rent away from the rest at the line of the eave *r r*, twisting and drawing out the cross lintels from their sockets in the walls of the house at the eaves, the whole mass then falling in the direction of *m s*. The weight of the timber-work was insignificant, the chief mass was in the heavy tiling which it carried. A line drawn from the position of the centre of gravity of this (the projecting tiled roof) to the centre of gravity of the mass of rubbish upon the ground, I found made an angle of 60° with the level ground. This is, therefore, about the angle of emergence; but as the roof received a small amount of nearly horizontal motion at its centre of gravity before it began to descend, this angle is a little too small, and so it may be taken as about 62° to 64°, coinciding as thus separately determined, with the emergence as given by the tiles projected from the tower.

The villa stands upon deep clay, upon the north side of the Auletta valley, the land sloping gradually southwards with a rolling surface, and behind it still further north, the limestone mountains (apparently Apennine limestone) rise with a flowing sweep to perhaps 1,500 feet, with bedding nearly parallel to the face of the slope, and a strike in the line of the axis of the valley, or E. and W. and north of E. The calcareous breccia is probably the rock directly under the clay on which it stands, however.

By barometer I find the ground at the villa is about 110 feet above the summit of Auletta. Returning from the villa to Auletta, a good view is afforded to the N. W. and W. down the great valley, which I sketched. Castelluccio is visible, perched upon its summit in the centre, and the lower ranges of the shaggy wooded precipices upon the north flank of Monte Alburno, block up the left of the view. Contursi is just visible upon another very distant height upon the right, (*c*) and the villa Carusso is seen at the roadside at (*b*). I proceeded on to

PERTOSA.—It also stands upon the top of a mound, less lofty and steep than Auletta, and not so elongated in form. The longer axis is on the whole transverse to the course of the Tanagro, or about 30° E. of N. The steep side,

FROM THE PLAIN OF PÆSTUM

FROM THE PLAIN OF PÆSTUM.

The flanks of MONTE ALBURNO & CASTELLUCCIO from near AULETTA.

faces towards Polla, or towards the S. and S.E., and the gentler slope towards Auletta, or N. and N.W. The steepest slope is about 55° from the vertical. The "colline" consists of solid beds of breccia of great thickness, of pebbles chiefly calcareous, but with many metamorphic hard slaty and chertose pebbles, of from 3 to 5 inches diameter, in a calcareous cement. Beneath the town these beds have a general E. and W. strike, or lengthwise to the river valley, and a dip of 30° to the S. and S. E., so as to be not very far from parallel to the faces of the limestone mountain slopes, behind them to the N. and N.W.

The town was an extremely poor place, the land of the commune being less productive than that of Auletta, and chiefly the residence of working peasants. With the exception of a few new houses, low, and tolerably well built, with dressed quoin stones and jamb linings in long blocks, which have stood pretty well, though heavily fissured, all the remainder of the town was built of large oblate or ovoid calcareous boulders, of from 10 to 12 inches across, picked out of the banks and river bed below, and laid into the walls as in Figs. 143 and 144 (Coll. Roy. Soc.), without an attempt at dressing flat-beds upon them, and with thick mortar joints. The town has hence suffered fearfully and is almost completely demolished. The timbers of many of the houses after their overthrow took fire, and more than 150 corpses of the vast number buried in the ruins here, were found charred and calcined when disinterred, and in some few cases, all semblance of humanity, from the action of the quicklime produced from the calcined limestone, absolutely obliterated.

It was very painful to witness the subdued and patient

endurance of the survivors, exposed to the terrible inclemency of the winter nights here, and the processions of women and children, carrying about some paltry relic, and whining litanies to the Madonna, in tones of despairing sadness. The recent visit as far as this spot of the deputation of Englishmen who came to distribute personally, the alms collected at Naples amongst our countrymen, was fresh and grateful in the memories of these poor beings, who crowded round me and struggled unwelcomely, to kiss my hands and pour " benedizioni " on all " Inglese " to such an extent, that had I not fortunately found the Padre, Vincenzio Mancini, I should have been unable to make any observation.

He was a man of much more than the average information and intelligence of his class, but conversed in no modern language except Italian, which was strongly provincial, and I found it difficult to follow him; he spoke Latin with some fluency and elegance, and I obtained from him a good deal of information.

The town has suffered most, at the east and west sides, the southern end the least. Large portions of the west side are a perfect chaos of ruin, and beyond reach of observation or analyzation, as in Photog. No. 146. The general character of the west side is seen in Photog. No. 147. The Photogs. No. 148 (Coll. Roy. Soc.) and No. 149 are views at the southern entrance near the town and towards its central part.

Those No. 150 (Coll. Roy. Soc.) and No. 151 give a tolerable notion of such portions of the town, as presented measureable elements of the direction of shock. The sketch No. 152, taken on the spot, shows a part of the west side in which the lines of street and houses were for

PERTOSA.

THE TENEMENTA DELLA MADONNA CAMPESTRINA.

a certain range nearly cardinal. The walls a and b, and all parallel to them, run nearly E. and W. These were in great part, more or less standing, but fissured, in sloping cracks; but the walls whose plane had been N. and S.

Fig. 152.

were universally down and in rubbish, unless something had accidentally propped them up. In the sketch No. 152, the gable wall a, had been built within about $2\frac{1}{2}$ feet S. of b, a passage had probably run between them; b was kept up by flooring and other props to the northward, but a had swayed over at top to the northward, and leaned against b there, the bottom of both being buried in rubbish wherever the fracture had occurred.

The horizontal force and velocity with which a had been brought over against b, was small, as they only touched, at the top, and the space below was all hollow between, and yet a was not fractured by the coming in contact: it was about 2 ft. 3 in. thick and about 15 feet high above the rubbish. The sockets of the joists, whence they had been drawn and twisted out, were still visible on its face.

The direction of wave-path given by all this portion of the ruined town, was from E. to W. 118° to 120° 30′ W. of N. The force in the N. and S. direction, therefore, which brought the wall a against b was necessarily small. In some other portions of this west side, the crooked streets brought the general position of the houses to be ordinal, nearly 40° W. on the average. The houses that were so circumstanced, were less absolutely demolished than those whose front and rear walls ran nearly cardinal or N. and S., but diagonal and crossing fissures were to be found in all directions, and huge wedge-shaped masses, as in sketch Fig. 144 were thrown out from the W. and S. W. quoins of numbers of them. From the extent to which destruction had proceeded, and the vile class of the masonry, measurements of the widths of fissures, were uncertain and the angles of fracture ill defined. From measurements and general judgment together, however, it appeared to me that the angle of emergence here, must have been extremely steep, steeper than the indications at the Villa Carusso, and certainly not less than 70° with the horizon, possibly as much as 75°.

This is corroborated by the nearly universal fall of the tiled roofs and heavier floors.

In the Photogs. Nos. 149 and 150, some of the steeply-inclined fracturing is visible: some of the temporary roofs, as well as the piling up of the stones seen in No. 150, had been work done since the shock, in digging out for interment, the buried corpses.

At the east side of the town, the wave-path was more nearly E. and W., and gave a general direction of 83° 30′ or 84° W. of N.

At the south portion of the town, the destruction was rather less than over the remainder. An obvious reason for the fact is afforded, now that the wave-path is obtained. The plane of the great breccia beds upon which the whole stands, is not very far from being at right angles to the direction of wave-path; hence the southern portion of the town received the blow through the *greatest thickness* of these beds, and thus, by the numerous and successive changes of medium, in passing from bed to bed, the force of shock here had sustained the largest amount of loss of *vis vivâ*.

The direction, of the longer axis of the hill on which the town was perched, also accounts for the greatest damage having been done at the E. and W. sides, and less upon the very top and more level portion of the place.

Nobody here nor at Auletta seems to have felt any second shock, occurring within a minute or two after the first; but they all speak of *a* second shock much less powerful than the first, but occurring about an hour after the first; testimony which appears strangely inconsistent with that universal at Naples. The Padre Mancini is positive, that there was no second shock soon (*i. e.* within a few minutes) after the first, of a noticeable character. He himself had escaped the first with difficulty, and only owing to the fact that he was not in bed, and so was able to rush out instantly.

As to the time of the shock, he does not know of any clocks having been stopped, inasmuch as there are none either here or at Auletta. He thinks it was at about a quarter past ten (*tempo Francesi*), *i. e.* not reckoning by Italian hours.

According to his narrative, the shock was from S. to N.; but when I caused him to point with his hand to what he deemed the north, he pointed nearly to the west, and I then found, what I afterwards recognized as the usual mode of speech in the provinces, that such words as "tramontana,—dal nordo,—meridionale,—mezzogiorno," were more commonly applied to the apparent path of the sun in the heavens, from rising to setting, than with precision as signifying points of the compass. I henceforward always made the narrator point out with his hand the azimuth he intended. The testimony of the Padre therefore, which coincides with that of the postmaster down at the bottom of the valley, agrees generally with the deductions from my observations.

Padre Mancini says, the first shock was "sussultorio," and immediately became "orizontale ed oscillatorio;" and he thinks the second, of an hour after, was "vorticoso." But upon being pressed as to what he meant, he at length said "he thought the direction was not the same as the first, and changed while yet shaking; but he was not certain," adding quaintly, that his own head and those of his parishioners had become "vorticosi" from the alarm of the first shock.

The accounts given here and at Auletta, of the sounds heard at the same time, or *rather before* the shock, agree in the main. At Auletta, those who were in the town and survive, commonly used such words as "fischio sospirante," and the like, in describing the sound. Those down at the Locanda generally said it was like the "romore di carozzo" simply. The Padre described it as "ronzio e romore moderato." And when asked more minutely to describe it, he said it was "ingens fremitus retonans cum sibilatione." My final impression was, that the description of the people in the bottom of the valley, conveyed the notion of a more confused sound, than that of those on the summits, in the two towns; as if echoes or secondary sounds from the precipitous sides of the hills around had in some way been heard more by the former than the latter, with the primary sounds.

Descending with Padre Mancini, from the town to the road at the valley bottom, which is perhaps 100 feet above the bed of the Tanagro close beneath, I examined the little Casa Communale, a nearly new building (about three years old) a parallelogram of two stories (Fig. 153), nearly

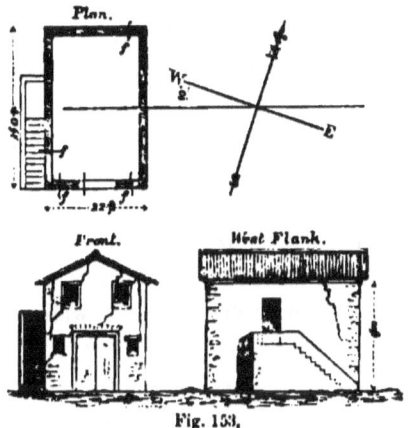

Fig. 153.

cardinal (axial line 10° W. of N.), built of good rubble, with cut stone jamb dressings, and middling quoin stones, and with a heavy external stone staircase on the west flank. It was fissured in several places, and the fractures gave a wave-path of 120° W. of N., and an emergence of about 65° with the horizon. The emergence might be steeper, however, as the direction of the fissures in the front, had been coerced in some degree, by the large proportionate surface, of cut stone jamb dressings.

CHAPTER VIII.

THE CALORE AND TANAGRO — ST. MICHAEL'S CAVE — GEOLOGY OF THE VALLEY — CAMPOSTRINA — GREAT ROCK FALLS.

ACCOMPANIED still by the Padre, I then crossed the river, and scrambled up the opposite bank, composed of deep clay and boulders, with gravel, to the mouth of St. Michael's Cave, from out of which issues that portion of the whole waters of the Calore, (as the Tanagro above this is called); that entering the limestone at Polla, at the upper end of the gorge of Campostrina, and finding its way by subterraneous channels, here debouches, and turning some primitive water-wheels at the Catalan forge, falls in a pretty succession of cascades to join the Tanagro again in its open bed. The mouth of this large cavern is probably from 50 to 80 feet above the open bed of the river opposite, and its jaws present evidence of powerful erosion. There are large stalactites within, and at about 300 feet from the mouth further entrance is barred by the water, which issues from the cavern almost clear and pellucid, while that of the river below is as white as milk, with impalpable cretaceous matter in suspension. As the water of the Calore thus divided between the subterraneous and the open channels at Polla is very muddy, and contains much calcareous sediment, it is obvious that that portion which passes through the subterraneous channel is filtered, or at least deposits much of its solid material on the passage, and is no doubt now forming new and strangely situated beds of clayey limestone, within the cavernous heart of the mountain range through which it passes, and which here separates, as by a huge wall, the valley of the Tanagro, from the Piano di Diano.

A short way within the cavern is a shrine of wood, with a rude plaster figure of St. Michael, of about three feet in height, in the interior. I found this 'genius loci' had been overthrown by the shock, and as the shrine is fastened up like a sort of cage, the figure was still leaning supine, against the back of the box at a slope of about 30° with the vertical. The saint had fallen, in a direction from W. towards the E. The base of the image projected widely in front, but less at the rear, and the figure being of a very upright character, it had been thrown over by the first movement, to an angle beyond the range of recovery, by the return stroke of the wave, and so remained out of the perpendicular.

The velocity of the shock had been sufficient to upset the image, but had not been sufficient to overturn the square wood shrine or cage in which it was placed, and which was about 8 feet high and $3\frac{1}{2}$ feet wide in the E. and W. direction. A sufficient corroborative proof, of the steepness of emergence of the wave here, is afforded by the stability of this shrine, near the top of which, was the stage on which the figure stood. An extremely small velocity, if horizontal, or nearly so, in direction, would have sufficed to overthrow the whole affair.

The volume of water delivered by the stream from the cavern, does not seem to be above one-twentieth that of the main open stream of the river below, which receives no tributary of importance between this and the other end of the subterranous duct, at Polla.

Close to this cavern, the northern end of the gorge commences, through which the open stream forces its way—a jagged, wall-sided cleft, betwixt precipices, of rather soft, ill-bedded, and cretaceous *looking* limestone, nearly white in fresh fracture, and whose mean height is probably about 800 feet.

Standing upon the left bank of the river, and 50 feet or so below the level of St. Michael's Cavern, one is enabled to see with some clearness, the geological relations of the main valley and of its lateral ranges, which in transverse section here (looking southward) are approximately shown in Fig. 154. The Apennine limestone of Monte Alburno and its associated range, dips to the south-west and south, but with a constantly varying angle of dip. What is the connection of the beds, of the opposite or eastern chain, with those at the bottom of the valley, is not traceable; it may be one of disunion and dislocation, as the existence of the gorge of Campostrina would suggest; but the beds look, upon the whole, to be parallel continuations at the east side, of those deep under Monte Alburno upon the west.

Above these in the valley, lies the coarse calcareous breccia, in beds approaching conformability at the east side; but where visible through the telescope, at the summits of the underlying ranges, on the west side, appear to be wholly unconformable to the escarpments of Alburno. Small but irregular valley bottoms, are formed at both sides, between

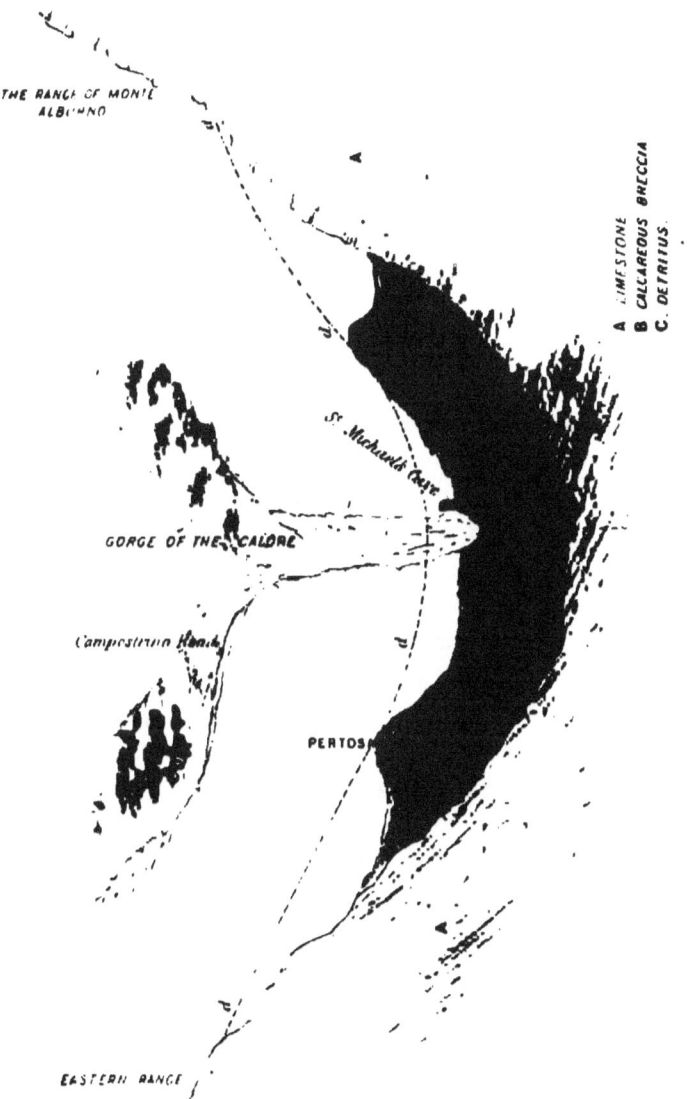

the underlying summits of the breccia mountains, and the flanking chains beyond; and in these, as well as in the bottom of the main valley, a great depth of loose material is deposited, chiefly heavy calcareous clays and boulders. These are deeply cut into by the lateral torrents, and still more deeply by the Tanagro itself, which here rolls over a bed wholly of rounded boulders, the skeleton of the washed-away detritus.

It is not easy here, or indeed anywhere else in the Southern Apennines, to imagine the train of causation (upon any of the usually accepted views of elevation) that led to this formation. It seems probable, however, that before this upper part of the valley assumed its present character, the surface of the breccia occupied something of the line $d\ d$, &c., if not one still higher, and that enormous masses have been removed by denudation, between those which now form the opposite ranges of "Collines." As respects our immediate subject, it will be obvious that any earthquake shock, emergent from the eastward at a steep angle, must arrive, through an immense thickness of beds of limestone first, and of breccia afterwards, before reaching the surface; and hence with vast loss of *vis vivâ*, and buffed, as to much of its destructive power.

I was unable to attempt determining, whether the breccia beds lie *directly* upon the limestone at both sides of the valley, or may have some other thin beds interposed. But I think the first is the fact.

The military road of Campostrina, over the rampart that separates the valleys of the Tanagro and of the Calore (as its higher stream now is called), is led over the mountain at the eastern side of the river gorge, winding round several lateral valley-gorges, and crossing the principal one by an imposing viaduct of considerable altitude, built of ashlar limestone, and carrying a narrow road, over a double range of semicircular arches, upon piers overloaded with material and buttressed out, transversely to the width of the road, to more than twice its breadth at their deepest bases. This viaduct must have received the shock very nearly transversely to its length, but emergent at a high angle. It has sustained no damage whatever, though a top-heavy mass, a sufficient proof of the value of good masonry in an earthquake country. Padre Mancini politely accompanied me on foot to the summit of the pass, about 2½ Italian miles, to point out the site, of an enormous fall of limestone rock, which had been produced by the shock.

The general direction of the deep narrow gorge, in the bottom of which the foaming torrent of the Tanagro rolls for several miles along, is nearly N. and S. The limestone beds at either side, as well as the jagged serratures of the cliffs, in many places vertical or overhanging, correspond to each other, and prove it to have been torn by separation of the opposite mountain masses.

The strike of the beds is nearly N. and S., and they dip at various angles, but all very steep towards the S. and S.W. The bedding is not very clearly defined, and the rock is lithologically, softer and more of a cretaceous character, than that of Monte Alburno, and may probably belong to a different member of the limestone formation. No fossils anywhere met my eye, and other occupation precluded my looking for them.

The first great fall of rock I found had occurred about

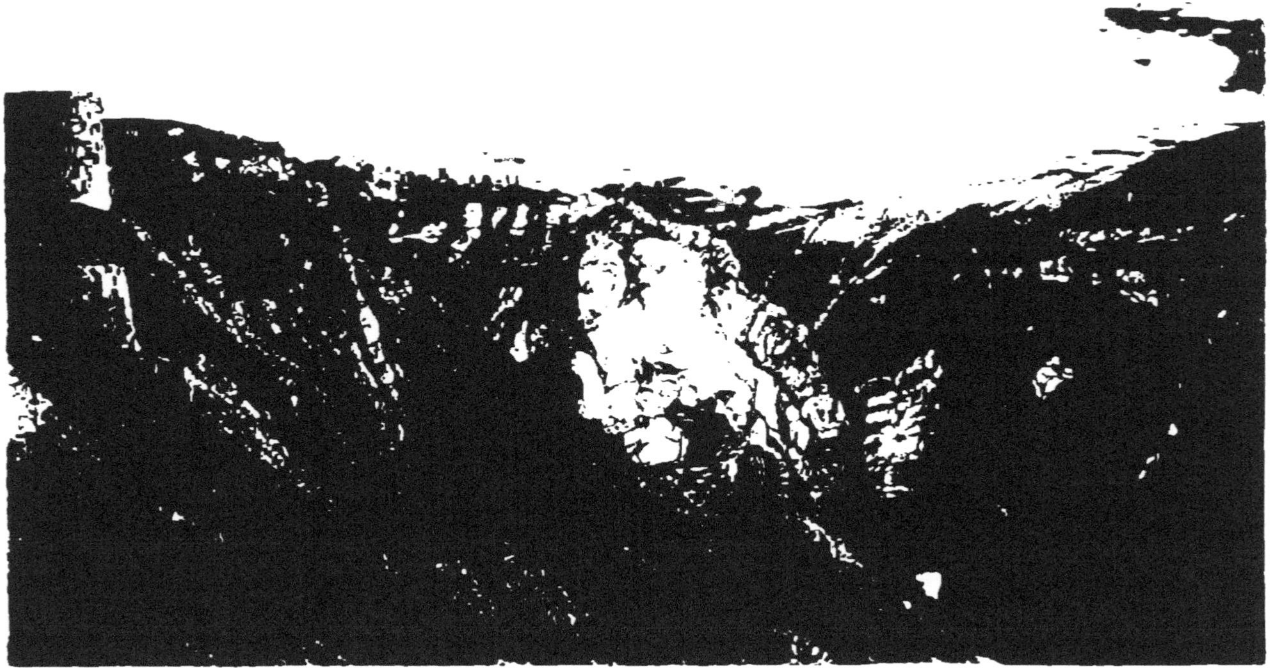

a quarter of a mile beyond the viaduct: its position is shown in Photog. No. 155, the view in which is taken, looking southwards, some of the revetment walls at the turn of the zigzags of the military road, mounting the hill, beyond the viaduct, being visible at the left of the picture. Near the centre of the view, where the white face of rock is visible, a mass of limestone had been thrown off in a S.W. direction, 69° E. of N., from the face of the cliff, and, shattering in its fall, had carried some thousands of tons of rock, small stone, and fine whitish debris, down into the bed of the river. The talus was heaped up at the base of the cliff, and stretched nearly across the whole breadth of the bed, forming a serious obstruction to the current, which was dammed back a good deal, notwithstanding the rapidity of its slope, and the torrent had already washed away large portions of the fallen mass, and as I watched it, was sorting out the finer material, and as it was removed slight falls of stuff continually occurred at the toe of the slope of debris, into the water rushing past its base.

The Padre informed me, that for four or five days after the 16th December, the volume of water discharged at Pertosa, both by the open channel of the river, and by the cavern of St. Michael, was visibly diminished, he thought by as much as one fourth the former delivery, and both currents ran turbid and foul. The delivery of the cavern, he thought, was even now, less than what it had been before, but the open river had returned to its usual regimen, except in so far as it was much whiter in colour, from the suspended chalky limestone, than he ever remembered it, though always more or less, so discoloured.

I remarked that many other smaller falls of rock had taken

place, at various points of the gorge, and from the N. E. side of the lateral ravine, which is crossed by the viaduct, whose steeply inclined sides are in many places, covered with loose material and large angular boulder blocks. Several of these had been dislodged, and projected into the bottom, leaving in some cases the torn traces of their headlong descent, in furrows whose direction I found to be, about N. 15° E., the stones falling to the southward.

On gaining the summit of the ridge, next above the viaduct, and looking to the S.W. across the gorge to the opposite mountain, I observed a very singular cavity in the slope of the flank, and at such a distance back from the edge of the cliff, as would render it probable it may be vertically over, the subterraneous duct in the rock, carrying the water from Polla to St. Michael's Cavern.

This is sketched in Fig. 156. It appears as if produced by the falling in of the roof, of a cavernous enlargement of the subterraneous duct at this point; and the mass standing up in the middle of the crater-like cavity, is probably part of the roof, tilted over in the fall, and sustained by other fragments beneath. The Padre said there was no water at the bottom, nor any entrance from it to a subterranean chamber, and it had received no alteration, that he was aware of, since the earthquake. The cavity is probably a quarter of a mile long, from right to left in the sketch.

Amongst the many lying wonders that were narrated about the earthquake, I afterwards heard it circumstantially affirmed that this, was a crater, had been *formed* at the time of the shock, and that fire had been seen to issue from it.

At nearly the highest point of the road, I found the

CAMPOSTRINA.

INTERIOR COURT, PALAZZO PALMIERI, POLLA.

ATENA

barom. 28·09 in., thermo. 42° (14th Feb.), which, when reduced, gives the elevation = 1913·4 feet above the sea. This proved soon after, however, not to be the very highest point, which I reached at about 150 feet higher. The total elevation, therefore, at the road is 2063·4 feet above the sea, and about 1420 feet above the piano of the valley bottom at Auletta.

At the former point nearly, the road is formed upon a side cutting and small embankment, on limestone covered with 3 or 4 feet average, of arable clay land. It is sustained by an ill-built revetment wall of dry stone, with mortared top courses, in all about 12 feet high. The general direction of the centre line of the road is 30° W. of N., and nearly level. For a length of about 300 yards, an irregular longitudinal fissure was open, in the surface of the roadway, at about a quarter of its breadth from the revetment wall, of about 3 to 4 inches in width, and in some spots still, 10 or 12 inches deep, though much obliterated and filled by rain washings.

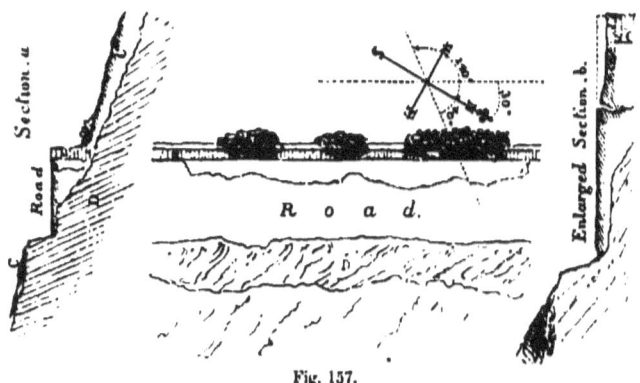

Fig. 137.

The form of the fissure and section of the road, as sketched, are shown in Fig. 157. Three portions of the revetment had been thrown towards the west, and from the position of the fallen material, the direction of the wave-path proved to have been N. 140° W. The portion of the road to the west of the fissure, had *slipped and descended* about 4 to 6 inches below the former level, as shown in section *b*. It was impossible to tell, whether the revetment wall had been founded upon the rock or not, but from the appearance of the ground at its base to the westward, I believed it had not: in any case, it had gone out at the base, towards the west, and with the mass of earth behind, had been severed from the remainder of the road filling, and slipped at the same moment. It is a case very analogous to the Auletta fissures, with this difference, that here, from the unsupported position of the road embankment and revetment, and the direction of the shock, the separation had much more nearly approached a "throw off" *at the instant* of shock, mixed with the movement of slip or descent.

About a mile further on, just before the rapid descent commences into the Valley of Diano, another set of road fissures had been formed, where the road is also in side cutting, but slopes off at the western side without any revetment. Here the fissures which are shown in Photog. No. 158 are clearly produced by the slippage off to the westward, of an enormous breadth of the clay land, reposing upon a surface of limestone—sloping westward at about 20° to the horizon—and with beds not much more inclined, and dipping in the same direction, circumstances all favourable to a large slip. The direction of wave-path shown, is about the same as the preceding. The telegraph poles all along this portion of the road, I remarked, had been loosened in the ground, and thrown out of plumb,

Prisoners on the Road near Pol(A).

and had not yet been again secured. Being restrained by the wires from rotating at the top, *i.e.*, confined to vibrate in a plane not very widely departing from transverse to the length of the wires, they had not formed conical cavities at their butts, but such as would have been produced by the pole, working forward and back, in a line not quite transverse to the length of wire, but, so far as its restraint would permit, also towards the south, so that on the whole their movement, coincided with the evidence of wave-path here given by everything else.

It was doubtless this swaying drag upon the wire (it is but a single one) produced by the poles that broke the former, and so cut off all telegraphic intelligence, between the great earthquake district and Naples, for above forty-eight hours, during which the most intense anxiety was felt in the capital, as to the fate that had probably overwhelmed the provinces.

Upon the highest summit of the pass is erected a little roadside shrine—the Capella della Madonna della Pieta, which was riven and fissured in a very remarkable manner, and only stood, by help of some pious props, that since the earthquake had been strutted against its tottering back and ends. It is shown in Sketch No. 159, made on the spot, and in Photog. No. 160 (Coll. Roy. Soc.), taken some weeks afterwards.

The plane of the front face of the "Tenementa," is north 45° W., and the fractures clearly indicated a wave-path having an azimuth direction of north 157° 30′ W., or from the N.N.E., and having a very steep angle of emergence. The little structure was built, of coarse limestone rubble, plastered all over, and the cohesion of the mortar joints but small. Applying to it the equation of overturning

$$V^2 = \tfrac{2}{3} g \sqrt{a^2 + b^2} \times \frac{1 - \cos \theta}{\cos^2 \theta}$$

with the value for V, (ascertained next day) for the emergent shock at Polla, it is certain that the split-off portions above b and c, would not have rocked and returned to their places as found, but have been completely overturned, had the wave-path been at a less angle to the horizon than 61° nearly. The angle of emergence here, therefore, must have been as steep as that, at least, and may have been steeper.

Looking back from this point, and sweeping the mountain side to the westward of the gorge with the telescope, I see a large Casale—upon the slope of the hill distant about four miles—almost in ruins, and can plainly discern by a favourable light, that it has been overthrown by a shock, which there, had the same general direction as here.

The direction of the slope of the hills at both sides towards the southern entrance to the gorge, begins to change and to trend round to the S.W.; and a little further on I catch the first view down upon the grand Vallone di Diano, its plateau level as a sea, stretching away twenty-six miles to the south, and four or five miles wide, the Calore, folded along upon its central surface like a silver cord, losing itself in distance, and the mountains rising almost abruptly from the piano at either side, the further end closed in and surrounded, by pile over pile, of dark grey mountains and snow-clad sierras, at last shutting out the horizon.

I can now perceive that Campostrina gorge, is the hinge,

between the two valley systems, and that the valley I have left, and that I am about to enter, have their respective axes almost at right angles to each other, the pivot round which they wheel being the mountain mass behind the town of Polla, and to the S.W. and W. of it. The descent now becomes rapid, and after another mile or so, Polla becomes completely visible, the dominant town of the north end, of the wealthy plain, along whose east and west sides I begin to discern many others.

Polla was an important place; originally, as its name imports, one of the ancient foundations of Magna Græcia. Nothing older than middle-age architecture remained, however, before the earthquake, and of this the Castello, near the summit of the town, was the most prominent. Its position in the rich country around, had produced its rapid modern growth to nearly seven thousand inhabitants, and most of its buildings were comparatively modern and pretty well built. Its streets and houses, churches and belfries, with olive yards and gardens between, spread themselves over the crown and slopes, to the north, south, and east of the large, low, short and well-buttressed spur of solid limestone rock, which juts out from the mountain range at the east side of the Vallone di Diano. The lengthway of this spur, is rather transverse to the general line of the valley, and its steepest side is towards the south. The city looked down upon the Calore, slowly and deeply sweeping past its eminence, and upon its own suburb of St. Pietro, at the opposite or right bank of the river, connected with the city by a fine old bridge of Roman style, and to the southward it gazed for miles over the glorious and unbroken hill-girt plain.

Its position and appearance are seen in Photog. No. 161. (*Vide* Frontispiece.) As I descended towards it, huge yawning gaps began to show themselves, upon the northern and southern slopes, where for acres in extent, everything had been levelled, all traces of streets annihilated, and where they had been immense mounds and sloping avalanches, of white and dusty stones and rubbish, filled up and encumbered the ground. Between these, shattered and bowing fragments of walls, and torn remnants of once lofty buildings, stood in mighty confusion; beams and rafters, tossed up like the arms of the despairing, stood out hard and black against the pallid heaps. The words of the Hebrew bard, referring to a still more eastern scene of earthquake energy, recurred to memory with a strange reality—" How is the city become an heap, the defenced city a ruin." Months of bombardment would not have produced the destruction, that the awful shudder of five seconds involved, when thirteen hundred houses fell together with deafening crash, and overwhelmed above two thousand of their sleeping inmates, and with clouds of suffocating dust, choked the cries of horror and anguish, that rose from the startled and often wounded survivors. In three different directions, conflagration soon added its terrors to the scene, and beamed up, a flickering and ominous light, into that dreadful night of cold and wailing, throughout the lingering hours of which, in helpless agony, they listened to the passionate entreaties for relief, the dying sobs, of relatives and friends entombed around them, and dreaded for them, more than for themselves, the recurrence of other shocks. The cold gray light of winter's dawn, obscure with smoke and dust, revealed hundreds bruised, or with broken limbs, without a roof to shelter them, many without a garment to cover them.

It required some hours' familiarity with such scenes, before the mind assumed sufficient composure and capability of abstracting the attention, to pursue the immediate objects of my inquiry.

CHAPTER IX.

THE OBSERVATIONS AT POLLA AND ITS NEIGHBOURHOOD.

Upon the south and north slopes of the town, I clambered over heaps of stone and rubbish, and amongst entangled beams, ten, fifteen, and even twenty feet in depth, above the former surface of the place. Upon the eastern slope the ruin had been less; but over the larger portion of the city upon the hill, the destruction had gone so far, that objects suitable for the determination of the precise direction of the shock no longer existed. Viewed with a comprehensive glance, it was obvious that the shock had been in a direction not far from north and south, and had been very steeply emergent. The Photogs. Nos. 162 and 163 (Coll. Roy. Soc.) convey some faint idea of the general appearance of the more completely overthrown portions of the city; and those Nos. 164 and 165 (Coll. Roy. Soc.) of the character of the ruins seen from the midst of them. No. 164 is a street of the more level part to the east of the city; No. 166 (Coll. Roy. Soc.) is amongst the heaps that overwhelmed the site where the church Santa Trinita had stood; and No. 165 looks upwards over a quarter of a mile of ruin, towards that of the Castello that

stood upon the summit, and under the masses of which many of the gendarmerie were crushed.

Wherever the walls had stood sufficiently, to observe the direction of fissures, they were found traversing the former at angles indicating steep emergence, as may be seen in the three last Photogs. In general, however, this angle approached the vertical more, at the upper parts of the city, than at or near the base of the hill on which it stood, proving that whatever had been the angle of emergence of the wave, at the base of the hill, or on the level of the plain, the hill itself, short and stumpy though it was, had vibrated with a proper motion of its own, which, being necessarily nearly horizontal, had thus modified the angles of the fissures. Descending by the east slope, and amongst the buildings around the base of the hill, and towards the bridge over the Calore, I found abundance of objects for fixing the direction of wave-path.

The church of the monastery of St^a. Chiara (Photog. No. 167) stands about half-way down the slope. The plane of the west end wall is exactly cardinal (north and south), and shows large diagonal fissures running quite across the façade. The direction of these proves a wave-path, north to south 10° to 12° E. of N., and an angle of emergence not less than 40°. The support of the solid belfry at the S.W. quoin, has prevented fissures crossing each other in this end, and, with the exception of one great diagonal shattering, at the top of the belfry, its construction, has involved chiefly a deep vertical fissure in it, extending almost to its base, but at the east end. The diagonal fissures of the main building, cross and have their upper ends inclined, chiefly towards the north. The roof has in great part fallen in, and such portions of it as still indicate the direction of throw, prove it to have been from S. to N., at an angle downwards of 50° to 60° with the horizon. Two small and slender iron crosses, one on the top of the pediment over the west end, and the other on the belfry top, are bent over, the former to the north to about 15° from vertical, the latter to the south, about 25° from same. I found I could not reach these, to get the scantlings of the iron, &c., as measures of velocity; the ruins continuing to fall at unexpected intervals, and the height being considerable. The belfry cross, by the theodolite telescope micrometer, appears to be about 2½ feet high, and of iron, about $1\frac{3}{8} \times \frac{3}{8}$ in. section, and is bent on the flat.

The church of the Madonna of Loretto was a solid Corinthian structure, still lower down, and in great part built of brick, with heavy semicircular arches, to the nave and aisles, and a heavy semicylindric roof; its axial line cardinal. As may be seen in Photog. No. 168, it is fissured down to its base, the fissures (some of which, of a regular and measurable class, may be observed under the altar-piece at the N.E. corner low down) all indicate a wave-path from north to south, and in direction about 160° 30′ W. of north, and an emergence of 50° to 60° with the horizon. The north flank wall, leans heavily out towards the north, as do all the large sashes still standing, high up in that wall. The main mass of the rubbish of the fallen roof, is in the inside of the church, and towards the north side of the floor. The shear or break of the roof vault also proves the direction of the force of fracture to have emerged from the north at a steep angle, as may be seen in the Photog. No. 168. On the right side near the pulpit, may be seen

a wooden Roman-Doric pillar, the base of which was fixed into the flagged floor, and which carries on its top a wood globe surmounted by a small gilded cross. The two last are bent over upon the iron spindle that confined both, towards the south in a plane very nearly north and south. This probably was produced by the second half vibration of the wave, but *may* have resulted from a blow from some falling body. The spindle or bolt that held ball and cross, was about $\frac{5}{8}$ in. diameter; the ball about 8 in. diameter; and the cross rose about 15 inches over it. I could not get its weight, but both it and the column were of hard wood. This church was built after the earthquake of 1652, and with special reference to future shocks. Iron chain bars had been built into the arches and roof, and would no doubt have done good service had the shock been more horizontal, but its direction of steep emergence took them transversely, and in several instances tore them across. The fractured ends of one, may be seen projecting from opposite sides, of the fallen roof-vault, and corroborate all other evidences of the steep emergence of the wave here.

Photog. No. 169 (Coll. Roy. Soc.) shows the interior of the church of St. Dominico (?), attached to a monastery a few hundred yards from the last, and looking westward. This was a much worse built church, and of stone. The direction of the inclined fissures may be seen, inclining northwards at their upper ends, in the west wall. The line of fracture of the arched groin of the roof at the south side, as well as the mass of the fallen material of the roof thrown towards the opposite or north wall, all prove the same general direction of path and steep emergence.

Photog. No. 170 presents an arched wall in a north and

south plane, 20° E. of north, in the same monastery, and an example, of heavy dislocated, and inclined fissures. The average angle here was 45° to 50° from the horizon. On principles explained in Part I., this is below the actual angle of emergence, as here each pier rocked more or less individually, and with the direction of the voussoir joints tended to give greater perpendicularity to the fissures. The crown of the upper tier of arches had all fallen in, and they, as well as the cross wall seen to the left of the picture broken off, had all been thrown northward, as the marks of abrasion upon the plaster of the face of the arched wall testify.

The Photog. Fig. 171 (Coll. Roy. Soc.) is an example of a frequent way in which arches are affected, indicating steep emergence. This arch is in a wall not far from north and south in plane. The ink lines $c\,c$ point out the extremes of minimum and maximum angles of emergence of the wave coming up from the northward. The dislocation of the arch produces loss of momentary support to the crown, the top ring of which has descended some inches, bringing a mass of wall along with it; but the fissure closing upon it at the second half vibration of the wave, has caught it, and the arch ring remains, nipped as shown, having descended through the space below its original intrados, at the point of intersection of the fissure, in rather less than the time occupied by one complete vibration of the wave. The data are not precise enough, however, to found calculation upon with any certainty as to this short interval of time.

The Photog. No. 172 (Coll. Roy. Soc.) looks N.W., and shows several tiled roofs fallen in, and is an example of the

SMALL HOUSE ON THE RIVER, PallA.
looking westward

HOUSE ON THE BANK OF THE RIVER, PallA.
looking eastward

general character of roof fall, here and wherever the angle of emergence is steep.

Descending still to the level of the plain, but not far from the hill of Polla, I found the most instructive examples.

The Photogs. Nos. 173 and 174, are views of an isolated sluice-house, that stands upon the east or left bank of the Calore, and was intended for use in relation to irrigation works, connecting the river channel with an artificial one at certain seasons.

In No. 173 the east side and north end of this structure are shown, and in No. 174 the west side and south end. The building is quite modern, constructed of good rubble masonry, with cut limestone quoins and jambs, flat window arches, and stone cornice over entablature. It is about 30 feet high above the soffit of the sluice arch, 30 feet wide north and south, and about 20 feet wide east and west, the walls 2 feet 6 inches thick. The building is exactly cardinal, its longer axis N. and S. Referring to No. 173, heavy inclined fissures will be seen running from both quoins, and meeting near the centre at top. The wall above the window arch is dislocated, and the voussoirs are thrown downwards, by a force emergent from the north, and nearly parallel with the fissure $c\,a$. Several minor fissures, which do not show in the Photog., existed through the arch joints near c, at 45° and 55° from the springing (or horizon), and in nearly the same direction, diagonally through the opposite or southern pier face.

The north end wall (to the right in Photog.) is scarcely fissured at all. The whole of the blocks of stone of the cornice at the north end, have been thrown from off the top, and falling not due north, but some points to the east of north, upon the brow of the bank below, have rolled down into the dry bed of the river, and are seen lying about under *c*. The cornice blocks tailed in upon the wall, but a short distance, and were not far from being balanced over the front arris of the entablature, on which they were bedded. The force that overthrew them, therefore, was one emergent from the north towards the south, a few points to the east of north; and from the position of the centre of gravity of the course of stones, it could not have been of steeper emergence than 70° to overthrow them, a line from that centre, at that angle, cutting the entablature arris. The roof had fallen almost completely in, and its debris of tiles and short timber, lay chiefly towards the north end of the interior.

Referring to Photog. No. 174 of the west side, where the wall was much more uniform, as to aperture, &c., the great line of fissure will be observed, inclining at top towards the north, and crossing through the arch lintel of the doorway about the centre. The angles of these fissures were 30° to 34° with the quoin, giving an emergence of 60° to 56° from the northward. The cornice was all perfect on the south end, as also on the east and west sides, except where fractured over the window at the east side, thus corroborating all the other proofs of steep emergence from the northward.

A number of the entablature stones, laid with wretchedly shallow beds, (only veneered upon the face of the rubble hearting,) on the centre part of this west side, are seen fallen out. They were lying very near the base of the wall, on the roadway that is seen here, passing over the sluice and in front of the building above it, and they had been slightly projected, by the eastern element of the emergent wave, at the moment when they were relieved from the detent of the cornice above, by the giving downwards, of the central part of the wall over the door lintel, which will be observed also down. The south end was very little fissured, and the evidence was clear of a wave-path, not varying above 15° or so from the plane of the side walls, or north and south.

The fissures were 2·25 inches wide at top, on the west side, and about 2 inches (in the same direction), on the east, which also indicates an eastern element of about 17° in the wave-path.

Referring back to Photog. No. 173, in the foreground certain piers are seen—groove piers for inserting stop gates, to pond the water for irrigation in the dry season. These stood in about 4 feet water, were of sound dressed limestone ashlar, and presented the long way of their horizontal section to the wave-path. Each pier was about 10 feet long by 4 wide, and about 12 feet high; not a stone was dislodged. The top courses were cramped together. From their peculiar narrow figure a very moderate divergence of the wave-path, from the north and south, would have probably sufficed to dislocate, if not to overthrow them.

Several other buildings of rather large size, situated about this portion of the city, chiefly ordinal, gave by fissure, when reduced, a wave-path varying between 157° 30′ and 164° W. of north. It would be unprofitably tedious to give the details at length.

One of the most instructive buildings at Polla, was the

Palazzo Palmieri, situated not far from the bridge over the Calore, and nearly on the level of the plain. The west front of the house (*i.e.* looking eastward) is seen in Photog. No. 176 (Coll. Roy. Soc.), the central building in the picture, and to the left of it is the Capella Palmieri, forming a connected building with the palazzo, and with its west end ranging with the front of the latter.

In Diagram No. 175, Fig. *a*, is a sketch elevation of this west front and block plan of the building. The house is exactly cardinal; it is large, comparatively new, and tolerably well built, though of short lumpy stone, with much cut limestone about the quoins and jambs, &c. It consists of a central court, surrounded by buildings of two stories and an attic, beneath the tiled roof. It is built upon the solid limestone rock, which at this point rises up through the deep alluvium, that lies around it at every side for a considerable distance. The ground is level around it except on the south, where it slopes off rather rapidly, the rock disappearing beneath the alluvium. It is, in fact, built upon the top of a sort of low pinnacle of limestone rock, that comes up (like many others round the borders of the great plain) to its level, through the deep alluvium, so that it stands as it were on the top of a little subterranean "colline," like a rock in an ocean of earth. The beds of the limestone here, as generally throughout the Vallone, are almost vertical, or at extremely sharp dip, and tend to a general east and west direction of strike.

The nearest rock, upon the same level, appears again at about 200 yards above the bridge, where are those singular and picturesque "swallow holes" into and through which, a portion of the waters of the Calore disappear, to

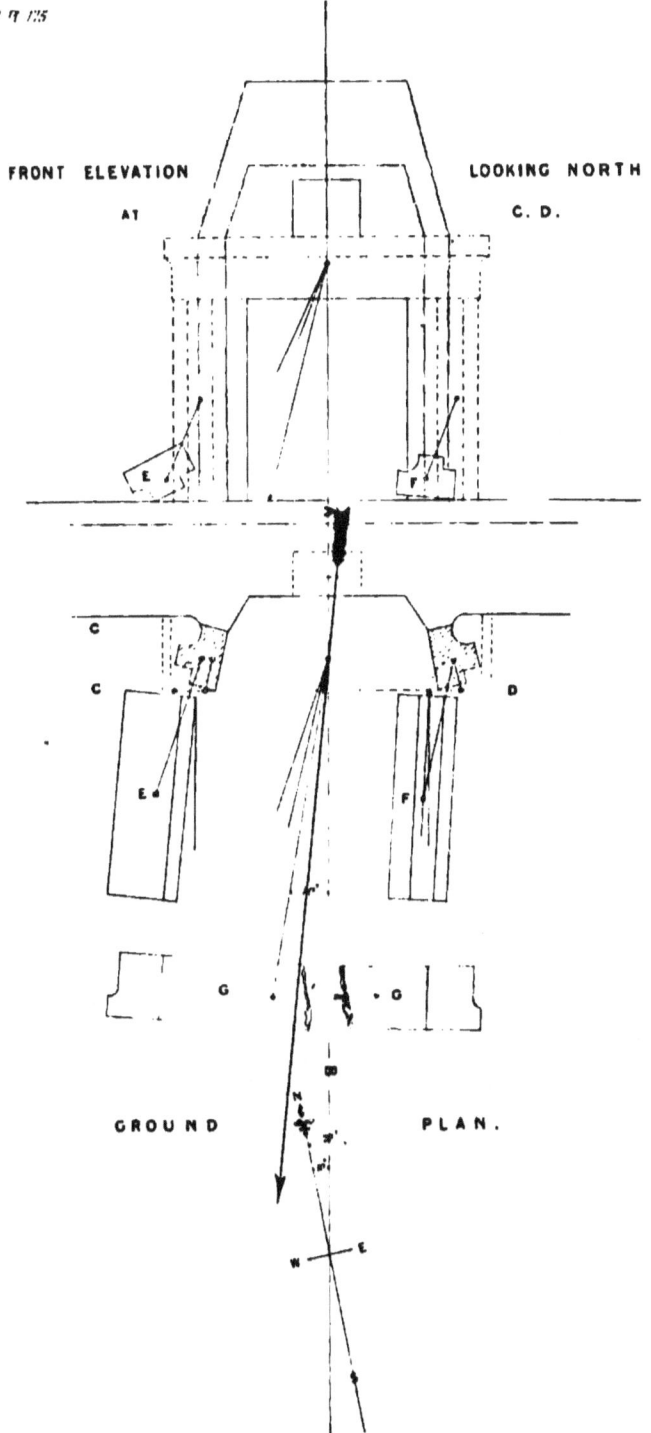

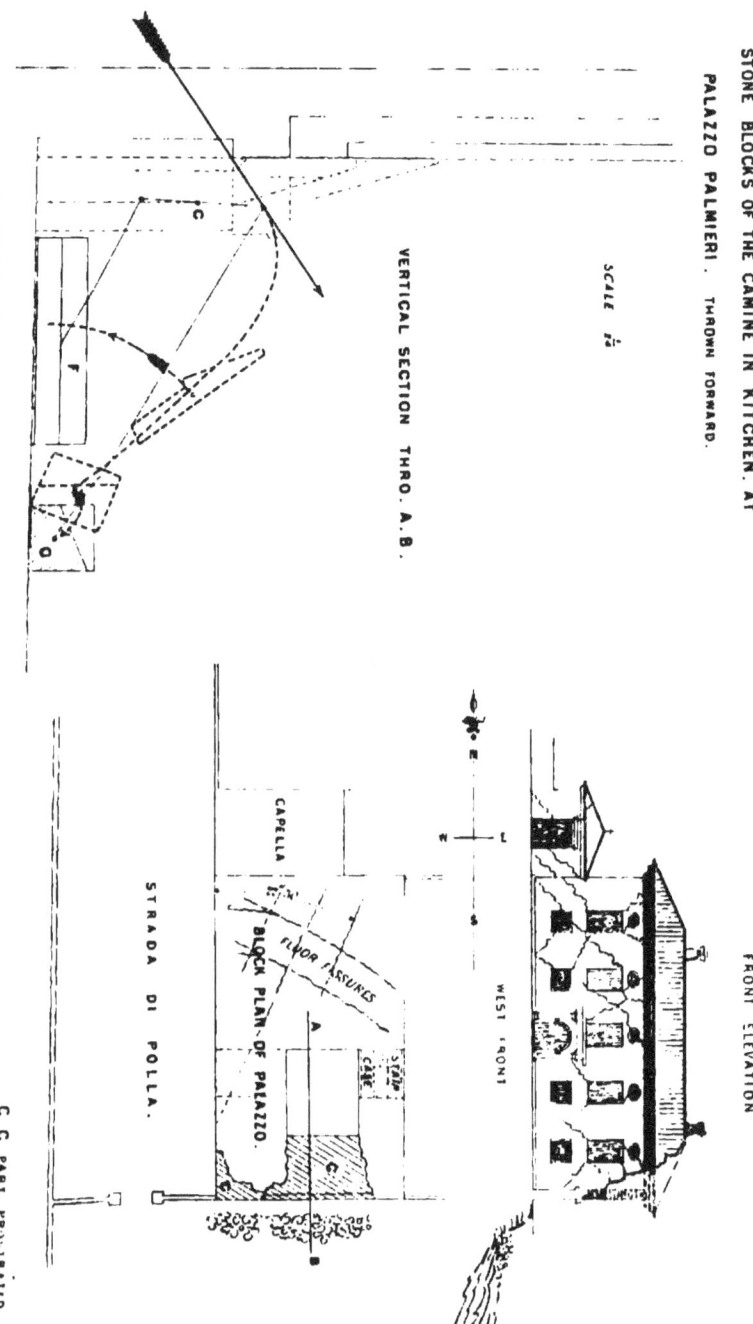

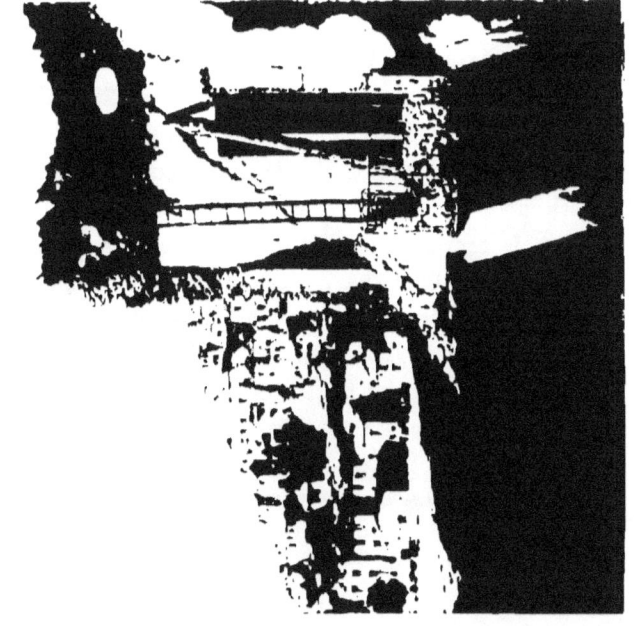

find their subterranean course to St. Michael's Cavern at Pertosa. There are several apertures, and the waters which before their plunge, turn some primitive old "molinas," appear to fall to a great depth. The rock, where visible, shows extremely rapid erosion by the water, as well as evidence of immense dislocation and denudation at former periods, when the great valley was drained dry over it, by the gradual rending of the gorge of Campostrina.

The Palazzo Palmieri is fissured diagonally in every wall more or less, those in the north and south walls being the most formidable. A large wedge-shaped mass, carrying with it a portion of the roof, is thrown from the S.W. quoin of the front, and a large portion of the south external wall, is prostrate and thrown to the south.

In Fig. 1, Diagram No. 175, the form, position, and angles, of the principal fissures found in the west front are shown, looking eastward, entering beneath the "Portone," and looking back or westward. The fissures formed above the archway in the north and south wall of the interior façade, parallel to the front, are seen in the Photog. No. 177, and in the sectional sketch (No. 178) taken on the line A B (on plan), and looking westward.

The south external wall, e to g, had been thrown to the south, and the upper part lay between t and k.

This is shown, in part, in Photog. No. 180 (Coll. Roy. Soc.), as seen from the balcony of the staircase, directly opposite the entrance gateway, and is shown in elevation in Photog. No. 179, looking eastward. In all these the general directions and angles of the fissures are correctly exhibited. There are some fissures, in all the walls, but the great mass are in those running north and south.

These make angles with the vertical of 30° to 40°. They cross in both directions, but the larger and greater number, are inclined at top towards the north.

The fissures in the walls running east and west are all more nearly vertical, than those in north and south walls. They prove that the building rocked in all directions, but that the main force was one emergent from the north towards the south, at an angle with the horizon of between 50° and 60°. The violence of movement in this direction was great, where, for example, one fissure passed through a joint of the arch, composed of long, curved, ill-formed voussoirs, at *s* (Fig. 178), over the front gateway, the arrises of the stones at the junction, are flushed off, as seen in enlarged Sketch (Fig. 178), and many such occur in the arches of the stone staircase, a square winding one, which, from its irregular form, is dislocated in every direction, as may be partly seen in Photog. No. 180 (Coll. Roy. Soc.) of the west interior façade.

The external wall *e g* at the south end, B on block plan, was built on scarped ground below the floor level. It had been about 38 feet in height, and was but little restrained by cross walls, and these only at the ends, about 70 feet apart, and was built of bad rubble (as is much of the Palazzo, though with costly cut-stone dressings, &c.), in lumpy blocks, of 10 to 16 inches greatest length, the mortar joints thick and bad, and no thorough bond. This wall was 2 feet 4 inches thick, and about 8 feet in height from the top, *e* to *g* (Fig. 178), was thrown outward, so that the great mass of the debris lies, in a parallel heap upon the sloping ground between *k* and *t*, from 10 to 18 feet from the foot of the wall, the average distance of throw being 14 feet, and the height from which it had descended 30 feet, *i.e.*, these being ordinates to the centre of gravity of the thrown mass. The portion thrown off from the top, was projected during the second semiphase of the wave, by the velocity impressed during the first, or by the return stroke of the shock.

Let $a = 30$ feet, $b = 14$. Then from the equation

$$\text{Tan } e = \frac{2H \pm \sqrt{4H(H+b) - a^2}}{a}$$

we find tan $e = 1\cdot593$, and $e = 57°\ 50'$, an angle of emergence which is within the limits given, by the fissures (which in this instance are extremely well defined in the Photogs. Nos. 177, 179, and 180, Coll. Roy. Soc.) in several parts of this building, and in that of the sluice-house at the river close by.

I shall recur to this wall when treating of the "camine," thrown in the dining-room.

Throughout the Palazzo, the floors are of beton and tiles, laid upon thick oak planking, crossing over oak and fir joists at about three feet apart: some in the largest rooms at the north wing, have been brought down altogether, the fractured beams showing, that they yielded to the inertia of the mass of beton and tiles, under the emergent wave. All are fissured, in various directions diagonally across, the lines more or less curved, with the *hollow sides of the curves towards the centre* of the floor, and upon the whole making angles of about 22° 30' with their respective walls, and crossing each other towards the mid length, nearly at right angles; the great prevailing direction 22° 30' E. of north, and at right angles to the same. These floor fissures, as nearly as they could be drawn by the eye, in one of the large reception rooms of the first floor, are shown in Fig. 181.

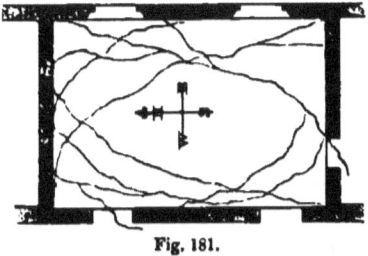

Fig. 181.

In the great drawing-room, the diagonal fissures in the north and south walls are 2, 2½, and 3 inches wide at the ceiling level, in about 16 feet in height. And the brown planks of the naked floor above, (naked here, because all the ceiling has come down, and left the oak and chestnut bare) have drawn from the north wall, which has gone out, and the floor and south wall have gone together in the opposite direction, the total movement at the floor level being 4½ inches, and almost exactly parallel to the line of the east and west walls. In another room, of 16 feet in length of end wall, the planking has drawn from the north end wall, 4 inches at the east, and but 2 inches at the west corner,

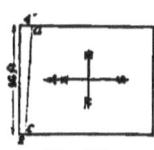

Fig. 182.

indicating a wave-path north and south, with some degrees trend to the east of north. The east and west walls being heavily fissured, the south wall had gone out in the direction parallel to ac (Fig. 182).

The Palazzo though thus shattered, and in many places dangerous to approach, had yet several articles of heavy furniture, &c., remaining untouched and unremoved; and through the politeness of one of the Palmieri family, who attended me with his servants, and answered all my inquiries, I was enabled to make a more minute examination of the interior of the house, and to record some most instructive cases of disturbance by the shock, giving measures at once of wave direction and of velocity.

Of these, one of the most valuable in deduction, is the overthrown stone and brick breast, of the "camine," or chimney hood, of the kitchen adjoining the great diningroom. This room was in the north wing of the building. The chimney hood was built, against the south side, of a wall running nearly east and west; so that a normal to its face, was north 10° E.

The chimney breast, which is accurately represented from careful measurements, in b, &c. Fig. 175, consisted of three blocks of Apennine limestone; two being vertical side jambs, and the third an imposited lintel, with a little brickwork superimposed, to complete the junction with the flue in the wall.

Fig b, 175 is the "camine," in front elevation, Fig. b, &c., 175 in ground plan, Fig. b, 175 in vertical mid section transverse to the line of wall. The dotted lines, show it as it stood before the earthquake, the hard lines representing the respective positions, in which I found the three blocks of stone lying, on the floor with the loose brick and plaster rubbish (chiefly) in the midst just as they had fallen; and Photog. No. 183 represents their appearance, and that of the wall from which they had been separated. The three blocks of stone merely *stood against*, the face of the chimney recess of the wall, and were made good to it with mortar, as was also the bit of pyramidal brickwork above the lintel. There was *no bonded connection between them*, nor any real cohesion, as the smoked surfaces of the faces of junction proved, that the expansion and contraction of the limestone blocks by change of temperature (when fires had been lighted) had completely severed the mortar union between them and the wall.

The whole fabric *stood*, therefore, *ready to be overthrown* by any force competent simply to overset it upon its base, in a southerly direction. The Figs. *b*, &c., 175 show that the whole mass had canted over to the south, turning upon the front lower arrises, of the side or jamb blocks, the latter carrying the lintel, for a short portion of the arc of descent, upon their upper ends. The brickwork above, owing to the position of its centre of gravity with reference to its base upon the lintel, when freed from the face of the wall, and the small velocity with which the mass must have been overturned, had fallen on the central space of floor, between the three stone blocks. The lintel block, had landed upon the floor, close to the upper ends of the side jambs, and had then, partly by the direction of descent, partly by the effects of its own elasticity, and that of the beton and tile floor, either fallen over, or slided (or both), forward a few inches further south, having been broken in two by the stroke on the floor, at a soft joint in the limestone. The two side jambs, from their peculiar form in horizontal section, and their setting, at an oblique angle to the face of the wall, had *canted round the inner angles* of the front arrises at their bases; and in their descent, the moment they got free from the lintel, had turned round partially, upon an axis parallel each to its own length, and lay, after thus rotating through about 90° each, with its east face uppermost, both having rotated towards the west; so that when fallen, E was removed further west from the centre line of the "camine," and F in the same direction *nearer* to the same line. The centre of gravity of the whole lintel block, in the same way was posited, at a point a little to the westward, of the vertical plane normal to the face of the wall, and passing through it as it had stood erect. The inner face (or cove) of the lintel, was uppermost as it lay on the ground, proving that it had simply been thrown down, without any secondary disturbing force.

Now this overthrow might have been produced either by shock from the south to the north acting upon the mass by its inertia, or by a projecting force, of a shock from north to south, carrying the wall along with it, forcing forward the chimney breast, and then at the return stroke, (or second half of the wave,) projecting it towards the south. Each assumption would lead to different results; but as there exists abundant and quite independent evidence that the general direction of wave movement at Polla was one from north to south, and *not* the contrary, we must conclude that this "camine," was projected, and thrown by the forward stroke of the wave. The rotation of the side jambs, and the position of final deposit of the lintel, prove that the direction was somewhat, from the east of north to the west of south, and geometrical considerations based on careful measurements on the spot, prove the wave-path to have been 165° W. of north which produced the overthrow.

Referring to Eq. IV., Part I., we can adopt the fall of the pieces of this "camine," as a means to determine, both the angle of emergence, of the wave-path here, and the velocity of the wave particle at its maximum.

The lintel G moving through a very small arc at the

commencement of its motion, may be assumed to have moved horizontally and free of restraint from the jamb stones. The height of its centre of gravity from the floor, was 4·33 feet; but inasmuch as the small bit of pyramidal brickwork above it, was separated and began to move along with it, we may consider the centre of gravity of the lintel raised by it to 5 feet. The block was thrown from this height to a horizontal distance of 7·2 feet, including that of its having once turned over, towards the south upon its front arris on reaching the floor.

We may further conclude, that *as it so came to rest, the velocity impressed, at the first moment of its motion in an horizontal direction, was not greater than sufficient to make it overset upon that edge.*

The difference of the side and diagonal of the lintel (in transverse section) = 0·62 feet; ⅔ of this is = H, the height due to the horizontal velocity of overthrow.

Then
$$b = 5 \text{ feet} . a = 7·2 \text{ feet},$$
and
$$-b = a \tan e - \frac{a^2}{4 \text{ H}}$$

from which putting in the values we obtain
$$\text{Tan } e = 1·472$$
and $e = 55° 49'$, the angle of emergence.

But $V^2 = 2 g \text{ H}$, the horizontal velocity; the total velocity, or that in the direction of the wave-path is therefore,
$$V^2 = 2 g \text{ H} . \sec^2 e$$
whence
$$V = 12·863 \text{ feet per second.}$$

If we apply this same method, to the *jamb stones alone*, or, as above, to the *lintel alone*, or to *the whole viewed as a single mass*, we arrive at the same value for V, within extremely narrow limits.

The assumption upon which this method depends (as respects the horizontal velocity impressed) is open to the objection of being slightly arbitrary; whether by compensation of errors however, or not, the result arrived at is extremely near to the truth, as will appear further on, and is controlled by the following calculation, which is open to no such objection.

Taking the overthrow of this "camine," in connection with that of the wall B, at the south end of the palazzo from which the upper part was thrown off. We have here two different bodies at the same spot, projected by the same shock, and by the same phase (the second) of the wave; and we can apply the method developed in Part I. (Eq. XL. to XLVI.) to determine both the emergence of the wave-path, and the velocity of the wave particle, in that path.

a and a', b and b' being the respective horizontal and vertical distances of projection, we have

$$-b = a \tan e - \frac{a^2}{4 \text{ H} \cos^2 e}$$

and

$$-b' = a' \tan e - \frac{a'^2}{4 \text{ H} \cos^2 e}$$

whence

$$\text{Tan } e = \frac{a^2 b' - a'^2 b}{a\, a'\, (a' - a)}$$

$$\text{and H} \cos^2 e = \frac{a\, a'\, (a' - a)}{4\, (a\, b' - a'\, b)}$$

and substituting for H its value $\frac{V^2}{2g}$

$$V = g \times \frac{a\,a'\,(a' - a)}{2 \cos^2 e\,(a\,b' - a'\,b)}$$

now we have for the "camine"

$a = 7\cdot 4$ feet, at the final point of repose
$b = 4\cdot 33$ feet

and for the south end wall

$a' = 14$ feet, $b' = 30$ feet.

Therefore

$$-4\cdot 33 = 7\cdot 4 \tan e - \frac{13\cdot 69}{H \cos^2 e}$$

$$-30 = 14 \tan e - \frac{49}{H \cos^2 e}$$

solving we find

Tan $e = 1\cdot 357$ and $e = 53° 37'$

and horizontal velocity $V_1 = 8\cdot 03$ feet per second and

$V = V_1 \sec e = 13\cdot 490$ feet per second.

The difference between the former calculation and this is $= 0\cdot 627$ feet per second, or little more than half a foot per second. If we take the mean of the two determinations we find

$V = 13\cdot 176$ feet per second.

There can be little doubt, that the small bit of brickwork imposed over the lintel block, of this "camine," in moving at first, along with the lintel, and with a longer radius at the commencement of the trajectory, communicated to the lintel block, a velocity slightly greater, than alone it would have been projected with, and that hence the velocity deduced from the measurements of the axes of a and b, may be slightly in excess. The true maximum velocity of the wave here therefore, as deduced from three independent sources, all corroborating each other, must be close to 13 feet per second.

A chimney hood of brickwork standing over the charcoal hearths in this kitchen, was at the end wall of nine inch brickwork A, (Fig. 185, Sketch Coll. Roy. Soc.) started out from the same line of wall, by about 2½ inches at top, in a height of 12 feet, but was not overthrown. It was *propped* however, by the return portion of the hood, at right angles to it, and so cannot be adopted for calculation, although affording a rude measure, that the velocity must have been small. In this same kitchen, the naked oak planking, of the ceiling or floor above, brown with wood smoke, shows where the ends of the boards have been *drawn* from their insertions in the walls, in the direction of their own length, and therefore transverse to the north and south joists, upon which they were laid, but which have not been moved, but to which the planks were not spiked or trenailed. The mark of the white mortar, shows the draw or shove to have been from north to south; it is 2¾ inches at the west end, and 4 inches at the east end, of the east and west wall, in a length of 25 feet; and as the normal to this wall bears 10° E. of north, the horizontal direction of wave-path deducible from this is as before, about 165° W. of north.

The actual amplitude of the wave in an horizontal direction cannot have greatly exceeded the average amount of shove, of these heavy beton and plank floors, and hence cannot have much exceeded 3 or 4 inches here.

In another room stands in a corner, against the wall T, which ranges north and south 10° E., and abutting upon the wall S at its southern end, a very heavy oaken household press (Fig. 184, Sketch, Coll. Roy. Soc.) It was 10 feet 6 inches high, 9 feet 6 inches long, and 2 feet 2 inches wide from front to rear. It was not thrown down by any east or west movement, but was shoved, by a north and south movement, along the beton floor upon its eight stumpy oaken feet, each 4 inches square (surface of contact with the floor), in a direction towards the south, $1\frac{3}{8}$ inch from the north wall against which it abutted.

The owner informed me, that it had been quite full of household articles, pretty uniformly distributed on its shelves; that it had not been filled so as to be top-heavy, or unequally loaded; and they estimated the total weight at about 800 rotuli = 1552 lbs. avoir. A quantity of china which occupied the upper third in height of its two southern divisions, was all found broken and thrown into a heap, at the south end of the shelves. A low velocity horizontally, not exceeding 2 feet per second, would have sufficed to make this press slide that distance upon a smooth floor, which would be only a little above $3\frac{1}{2}$ feet per second in the wave-path. It must therefore have been arrested by some inequality in the floor, or by its feet ploughing into the beton, and will give no certain result, except that the velocity was sufficient, to dash the contained china first north by the first phase, and then south by the second phase of the wave.

In one of the great drawing-rooms there stands still *in situ*, a very ponderous cabinet or chest of drawers of walnut (Fig. 186, Sketch Coll. Roy. Soc.), with its back against a wall running east and west, and free to fall to the south. It was quite full in every part with house linen, and estimated to weigh 400 rotuli = 776 lbs. avoir. (cabinet and contents). It rests on four irregularly-shaped feet formed by perforations in the front, ends, and back, and was neither overthrown nor shoved out from the wall at back.

Referring to the Equation V., Part I.,

$$V^2 = \tfrac{4}{3} g \sqrt{a^2 + b^2} \times \left(\frac{1 - \cos \theta}{\cos^2 \theta} \right)$$

we have here $\theta = 33°$, $a = 4\cdot75$ feet, and $b = 2\cdot80$ feet, as the mass must have fallen over, by rotating upon the most advanced angle at the base. Solving for V, therefore, we find the horizontal velocity necessary to have produced overthrow to be

$$V = 7\cdot378 \text{ per second.}$$

but e, the angle of emergence here, we have found to be 55° 49' from the "camine," and V, the velocity in the path of the wave, is

$$\mathrm{V} = V \sec e,$$

whence

$$\mathrm{V} = 13\cdot1328 \text{ feet per second.}$$

The velocity, V, given by the "camine," which *was* overthrown, is 12·863 feet per second by our first calculation; that necessary to overthrow this cabinet, which was *not* overthrown, we find is 13·133. We have therefore obtained the true value for V, within limit of the difference, or within 0·270 feet per second of the total velocity—a possible error of less than 2¼ inches per second.

The maximum velocity of the wave vibration at Polla may therefore be concluded with confidence to have been in round numbers 13 feet per second. The mean velocity obtained from both calculations, 13·176 feet per second, is a little more (0·043 feet per second) than sufficient to have overturned this cabinet; but the slightest inequality in its loading, or in its parts, by altering the height of the centre of gravity, would have sufficed to prevent its overturn, unless by a velocity greater than that assigned, by more than the difference as above.

Everything at Polla consentingly proved a wave-path of steep emergence, and from the north, or very nearly. The result, united with my observations in the upper part of the valley of the Tanagro, proved that the wave-path, as determined at Naples, either belonged to a separate focus altogether, or must have some complicated relation of disturbance, by reflection or otherwise, with the focus I was looking for, and which it now began to appear, I should find somewhere to the north, and not very far east or west of the meridian I was then upon. I therefore dismissed from my thoughts for the time the irreconcileable phenomena at Naples, leaving it to further observation to solve the apparent enigma.

The Judice of Polla, Signore Ferdinando Ganuzzi, politely accompanied me over the place. He had been at Polla on the night of the earthquake. According to his statement, confirmed by that of the Syndic, and others of the town, they were suddenly alarmed by the rushing sound, " Mormorio buccinante rapidamente;" and almost instantly, while it yet was heard, the first great shock came—" sussultatorio," succeeded in a very few seconds afterwards—how much they could not say, probably 10" to 20"—by another movement, that some of them habitually spoke of as a second shock, which was " undulatorio." But there was no such distinct interval of quiet, and arrival of a second shock, both being oscillatory, as was felt at Naples. The Syndic was on the second floor of his own house, which did not fall immediately; the first movement was sufficient to cause him to lose his balance when standing, and to fall upon the floor. He found some difficulty, owing to the second undulatory movement, in regaining his feet to fly from the shattered house.

No correct observations were made as to the time of the shock: all was confusion and alarm. He noted his watch, and by it the time was $10^h\ 15^m$ solar time, in Frankish hours. But he admits that all their watches and clocks are regulated by the setting sun, and are not reliable within narrow limits.

There was a second shock of considerable violence about an hour after the preceding, which shook down many buildings, that had been shattered by the first.

The Capo D'Urbano, a very intelligent man, made a curious and probably not unimportant remark, as to his experience of the sound. It seemed to him to reach him "*through his legs*, as he stood up," although, he added, "it was everywhere." This suggests the probability, that much of the sound in earthquakes may reach the auditory nerves, by transmitted vibration from the ground or other solid objects, through the bony skeleton; just as when a poker held by a string to the ear is struck, and thus may convey from a *very small vibration* an overpowering sense of sound to the auditory nerves.

I took from the centre of the middle arch of Polla Bridge, several intersections to correct magnetic declination by.

Monte Corticata bears	170° W.	of north.
Tower of Diano (highest tower)	179° W.	,,
Atena (high tower)	156° E.	,,
La Sala	152° E.	,,
Pizzo di Cirazzo	179° E.	,,

These give a declination varying between 13° 30' and 15° 30' west. I also took an observation of the sun, but on computing I find that some decisive error must have occurred in my note of the sun's azimuth, which renders the observation valueless, as there was no ground for supposing any material disturbance of the ordinary amount of declination at this spot, more than at La Duchessa, where the same error appears to have been made.

As I left Polla in the afternoon, the grand plain of the Vallone di Diano opened before me, level almost as a sea, of deep rich alluvial clays, which, as they approach the roots of the mountains that rise almost abruptly from the plain, assumes the form of a sort of sloping and almost continuous terrace all round the plain, elevated everywhere a few feet above its average surface, seldom rising more than 40 feet above it, frequently not half that.

The outline section of this terrace, as it sweeps towards the centre of the plain, is that of a flat, hollow, parabola-like curve, with a slope varying from 15 to 1 to 60 to 1, or even still more gradual, and suggests at once to the eye, the geologic conditions that produced the piano; once a large lake, or arm of the sea, which found egress for its waters at a level probably not lower than the summits of the Campostrina Pass. In that condition of things, the rich bed of mud was deposited, that now forms the basis of the valley and of its agricultural wealth.

The great fracture, through which the Calore now finds its course *sub dio*, must have been subsequently formed, and through it the lake was drained down to the level of its marginal terrace. Its subsequent progress of desiccation must have been gradual, and dependent upon the rate of erosive deepening of the river at its north extremity; and for ages it must have remained a shallow and probably pestilential lake, of above 20 miles in length and 4 or 5 miles wide. The total difference of level now, between the water of the Calore at Polla and at the junction of the Peglia with it at the extreme southern end, is said not to exceed 6 or 7 feet by Signor Palmieri of the Corps of Strade e Ponti, or only 3 inches to the mile. The marginal terrace round the plain contains numerous fragments of limestone, some angular, many more or less rounded. But the great clay central bed scarcely presents a pebble until high up towards the southern end of the valley, where the washings of the hill-side torrents disclose coarse gravel and boulders also, embedded in it. On the west side of the valley, I pass St. Arsenio and St. Pietro, both low-lying villages, placed upon the level of the marginal slope, that like a great shallow saucer, surrounds the plain. These towns have suffered but slightly, although not five miles in a right line from Polla, and St. Pietro, the more distant, has suffered the most. The damage done, however, is almost confined to old houses, built of the usual sort of wretched, short, nobbly, bondless limestone rubble, of rounded lumps like irregular loaves, of from 6 to 16 inches diameter, with thick joints, of bad mortar, made of clay rather than sand; the general direction of wave-path from 142° to 144° W. of north.

The character of the limestone of the mountains at both sides of the valley, begins rapidly to change, from that of the valley of the Tanagro and Salaris. It is no longer the hard, sharply-fracturing, clearly-bedded, and well-defined stone of the latter, but a loose, irregular, ill-defined, and *inarticulately*-bedded, crumbly stuff, extremely like in lithological character, the white and variegated limestone beds, of great portions of the upper limestone of Roscommon, Leitrim, and King's County, in Ireland, but much more sandy and siliceous.

Such bedding as is traceable upon the surface, appears in a direction almost vertical, and although with many changes of strike, having, upon the whole, directions ranging transverse to the main axis of the Vallone; and hence, upon the whole, presenting the flat of the bedding, to the direction of the wave-path along the valley. There is no general indication of continuity of bedding or of formation, at the opposite sides of the valley, nor anything that decisively points out, whether this limestone of its flanks continues right across beneath its clay filling or not. It seems highly probable, from the general structure of the country, however, that it does so; and equally so, that limestone, breccia, and marl beds, may lie between the crumbly limestone and the alluvial clays. The more distant and higher summits, visible clearly by the telescope—particularly those on the east behind Atena—present more of the character of the hard limestone of the Tanagro. The limestone of the lateral hills is so ill compacted, and so easily acted on by air, water, and carbonic acid, that it is cavernous in every direction, and weathers into holes and pits of fantastic forms. In hand specimens, its structure frequently shows it to consist of a mass of compacted crumbs, of rather harder limestone, angular, and not much unlike in size and form, Vesuvian "lapilli," but very slightly coherent by a softer calcareous paste.

This form, gradually passes into an almost white cretaceous-looking limestone, very soft, friable, and filled with disseminated fine white sand, in many places, particularly along the east flank of the Vallone. The hill sides everywhere present strong evidences of denudation, down to within three or four hundred feet of the level of the plain. Their summits and flanks are all swept almost bare, and the great deposits of detrital material above the plain, are only found in hollows, where it was entrapped, or in steep banks not elevated far above it, and now rapidly eroding by torrents. Almost all the scattered buildings that I pass along the plain, going southward, present indications of a general wave-path from north to south, varying some points east or west of that. On the east side of the great military road, however, about a mile from Polla, are some isolated buildings, not more than 150 feet above the plain, yet upon the solid limestone, which present decided east and west characteristics of wave-path—one especially, a nearly new building, having a wall of about 50 feet in length running north and south, and connected with others nowhere but at its extreme ends, by two transverse, or east and west walls. About 7 feet in height is thrown off the top of the north and south wall for its whole length, and thrown towards the west and south, while the east and west, or end walls, are merely fissured slightly. The original height of the fallen wall was about 35 feet, its thickness 2 feet, and the mortar was not yet indurated. The debris had rolled down a slope of 1 in 6, and much of it was 20 feet from the base of the wall, and hence did not admit of any calculation as to angle of emergence.

It was obvious that this was some local disturbance connected with the rocky range behind (to the east), but what I could not at the time discover. At about 2 miles south of Polla, and at a point of the road about the general level of the central plain here, I took by barometer the level of the Vallone. At 1^h 20 Greenwich time (14th Feb.) barom. reads 29·6 inches, thermo. 62°, which, reduced, gives 581·5 feet for the level of the central plain above the sea. Either by misreading, or through some great inequality of atmospheric pressure on this day, between Naples and the Vallone, this observation is certainly about 100 feet below the truth; for the bed of the Tanagro at Auletta was ascertained to be 576 feet above the sea. The waters, issuing slowly from St. Michael's cavern, are about 80 feet above the river beneath, and there is probably 20 feet fall between Pertosa and Auletta; while it is probable that the fall in the rocky tube is but a few inches between the water at Polla and its issue at the cavern. There must therefore be a fall between Polla and Auletta of about a hundred feet. We may therefore take the mean level of the Vallone di Diano at about 700 feet above the sea, and the barrier of Campostrino, before its fracture, stood 1,363 feet above the plain. This was the depth of the ancient lake or arm of the sea.

While at Auletta, the gendarmes and others there, had affirmed that on the night of the earthquake, they had seen some unusual sort of light in the sky or air, and some of them said that it appeared to come up out of the earth. At the moment, I looked upon this as a superstitious tale; the rather, as Padre Mancini had not remarked it; but I find almost every one I converse with in these districts, speaks of having seen some unusual halo-like light in the sky, before and not long before, the shock. Many describe it at second hand, and these differ much in their statements as to time of its appearance, and give no intelligible account of its character; but many others say they saw it, and attempt to describe it as something like the light long after sunset, streaming up from the horizon at some one point, a sort of zodiacal light or phosphorescent diffused halo; whilst some point to one direction, some another, as its azimuth of apparition. Others — and amongst these were the gendarmes at Polla — say that the light seemed to them to emanate from the earth itself; and those that were in the dark gloomy lanes of the towns before the shock, say that some sort of luminosity lighted them upon their way.

Much, most of this, may be but the fancies of an imaginative and wonder-loving people; but in a country where communication is so bad, and news travels so slowly, it would be remarkable if so widely-diffused a notion, and one without any obvious popular basis of suggestion, should be devoid of all foundation in fact. I have therefore recorded it; and before dismissing the subject I may add, that I found the same story prevalent in the valley of Viggiano also, but lost all trace of it farther south, east, or north, while to the westward I heard the first of it at Auletta. It therefore has this remarkable attendant circumstance that, if fabulous, the fable was confined to an oval district around the most disturbed region. Conjectures would be useless as to its nature, but future observation directed to the point, may determine whether some sort of auroral light may emanate from the vast depths of rock formation, under the enormous tensions and compressions, that must precede the final crash and rupture that produces the shock, or whether volcanic action going on in the unseen depths below may give rise to powerful disturbances of electric equilibrium, and hence to the development of light; just as from volcanic mountains in eruption lightnings continually flash, from the huge volumes of steam and floating ashes above the crater.

Atena, which in Pliny's time gave its name to the whole valley (campi Atenati), is a town of extreme antiquity, its name indicating its Greek foundation. It is situated upon the east side of the great valley, and stands upon the crest, of a spur of absolutely bare rock, jutting from the lateral range, and having a small transverse valley or gorge to the north of it. The slope or angle of emergence of the rock, to the north of the town, from beneath the deep alluvium is about 30° with the horizon at *a* (Fig. 187); the bedding ill-defined but apparently with a general east and west strike and steep dip.

Don Vincenzio Jachetti, an inhabitant, an intelligent man, who had been appointed a Deputy Sotto Intendente since the earthquake, accompanied me over the town, which has suffered terribly; its streets, especially upon the north and south sides, are choked with rubbish of fallen buildings to a depth of 10 or 15 feet above the former level, and encumbered with fractured and entangled beams and joists.

The Photog. No. 189 (Coll. Roy. Soc.) shows the east end of the cathedral church, and various adjoining buildings, looking nearly westward. The general evidence of the wave-path with steep emergence from the north, is observable in the thrown and shattered north and south walls, and in the wedge-like mass, projected from the S. E. quoin of the chancel, and the roof of it and of the south transept fallen *inwards*. The horizontal direction of wave-path deduced from the thrown quoin, and from the average of five sets of fissures in the church, was 165° 30′ E. of north, and the angle of emergence 47° 30′. The angle of emergence given by thrown quoins in three other buildings (dwelling-houses), of much worse masonry, however, and therefore capable of less exact determination, was only 39° 30′. Lower down upon the eastern slope of the town a large heavily built range of building, with front walls to the narrow street 3½ feet in thickness, running nearly east and west, shows fissures of two inches wide, on the level of the first floor above ground, which indicate a wave-path not far from north and south, 177° 50′ E., the nearest approximation. One of these in the soffit of the arch, over a " portone " leading to a " vicinello " is shown in Sketch Fig. 190, and gave a path exactly north to south. A wall of old and little coherent masonry, has had six feet in height of its upper part thrown off, parting horizontally along, at 38 feet from the base. The mass of material has fallen at an average distance of 20 feet from the base, towards the south, and a point or two towards the west. The angle of emergence deduced from this (the horizontal velocity as at Polla) is between 45° and 50°, dependent upon the constant adopted for the adherence of the mortar and masonry at the base of separation. The result sufficiently corroborates those above given from more certain data.

The difference of effect of the same shock, upon well and ill-constructed buildings, is forcibly shown here. The square campanile of the church stands nearly isolated, its north and south side walls being nearly parallel with the axial line of the cathedral; the east and west walls stand 120° west of north. It is about 90 feet in height and 22 feet square at the base. The walls, 3 feet 8 inches thick at bottom, and only 12 inches at the summit (Photog. No. 188½ and Fig. 180) are very well built, with large, long-bedded, heavy ashlar quoin stones, 3 to 4 feet bed along the face and 16 to 24 inches deep; cut limestone jamb linings and string courses; and the filling in between these, well-laid coursed rubble. At each of two points of its height—viz. the first and second string courses—the walls are connected by four slender chain bars of 1½ in. × ¾ in. iron, with transverse cotters outside the wall faces. This has stood uninjured, without even a crack, in the midst of surrounding ruin, a clear proof of what sound and good building would do, in securing the safety of the inhabitants of the towns, in earthquake countries. High up upon the rocky hill side above the town also, are many summer lodges (*scaffæ*) which are very well built, and of recent date; and although probably a thousand feet above the town level, they have suffered very little: they are chiefly buildings of a single story, and owe their safety to this and to their good construction.

A large portion of the ancient walls of the town remain, probably of mediæval construction. At one part of these a large cylindrical tower existed, which for ages had been used as a cemetery. From the side of this, overhanging the precipitous face of the hill, a large mass had been thrown, and had exposed to view, the surface of a solid cylinder of human bones, of several feet in depth, those at the bottom reduced almost to crumbled bone-earth, while those on the surface at top, were still perfect, and some not quite

ATENA.

A STREET IN ATENA.

denuded of ligaments; a proof how ancient in Southern Italy this barbarous mode of naked interment of the poor, (which is still in use at Naples) has been. The chief interest to science, however, lay in this; many of the bones and some skulls had been thrown from the mass along with the debris of the wall; upon the precipitous limestone slope where they rested, some small calcareous springs oozed out, and their deposited tufa was visible. It is not improbable that these human bones may become incased in tufa, and the latter may hereafter form at this spot a coarse conglomerate, with the fallen masonry and embedded bones.

The position of the tower is imperfectly seen in Photog. No. 191 (Coll. Roy. Soc.), and its appearance in Sketch Fig. 192 (Coll. Roy. Soc.): the tower was about 28 feet in internal diameter.

The time of the first great shock was marked here by the stoppage of the communal clock at $10^h\ 15^m$ Italian time reduced to Frankish, but no exact reliance can be placed upon this. Signor Jachetti admits that all their clocks and watches about the country are either set by sundown or by the watches of travellers coming from Naples or elsewhere.

Upon the bank of the Calore, out in the centre of the plain opposite Atena, Jachetti pointed out to me, a fissure in the deep clay soil, which had been opened nearly parallel to the stream. It was simply a land-slip of a few hundred feet in length, the fissure 6 or 7 inches wide, and the vertical descent about the same, and originated at the violent shake at the shock.

Immediately behind the town, in the small lateral valley,

a fissure also exists in the earth, which Signor Jachetti affirms when first opened extended into the rock beneath, but that the rains have since filled the latter in. I have much doubt of the fact, however, from his description of the appearance of the rock at the alleged fissure, which rather seems to have been an ill-defined junction of bedding at a steep angle, and that the outer bed of rock had slipped a little downward and outwards, over and from that on which it reposed. The earth fissure, however, was traceable for several hundred yards beyond the saddle-back of a colline, connecting the spur of Atena with the main range of lower hills behind. It appeared originally to have been about 2 or 2½ inches wide, and one side had descended about 3 inches below its original position. It was in earth over the limestone, varying from 2 to 4 feet in depth — a manifest case, like that in the plain, of slippage and shaking off of loose material, and not of actual fracture by bending or dislocation (Fig. 193).

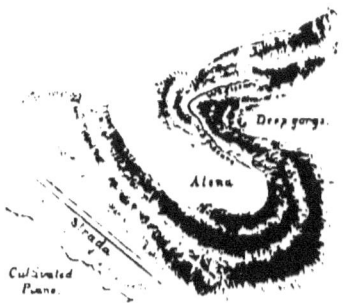

Fig. 193.

A second small gorge, with deep and precipitous sides, runs in an east and west direction behind Atena, and falls of crumbly limestone rock have taken place, from both its faces, but chiefly from that on the south side, where the detached masses have fallen from the ends of nearly vertical ill-defined beds, whose strike is N. W. and S. E. This, like every case of fallen rock that I have so far observed, has been detached from a vertical or nearly vertical bed, where, owing either to the joints of the beds themselves or to cross fissures, there was little or no adherent connection with the adjacent rock; in fact, cases of loss of equilibrium and fall by inertia, and not of rending asunder through the solid stone, and dislocation by the direct energy of the shock.

Upon the opposite or west flank of the Vallone, and further south, stands Diano, the town from which it takes its name, upon a low, stumpy, jutting-out spur, of soft limestone, to the eastward of the great range (see Photog. page 165, No. 108, Part I).

The main direction of this spur, is nearly due north and south by compass, rising gradually from the plain at the north end; and it is completely cut off from contact with the great lateral chain of mountain, except at nearly the level of the plain, by the long lateral Vallone del Raccio, which brings in one of the great feeders to the Calore on its left bank, and whose bed in the bottom of the vallone, seems to lie in the line of a great dislocation.

This town has suffered comparatively little by the shock, many fissures, and a few of the old ill-built miserable class of houses thrown down, direction apparently north to south, none affording good data.

The beds of limestone rock at both sides of the valley, from above Atena southward to below Diano on the west, and La Sala on the east, are vertical or nearly vertical (so far as bedding can be discerned at all), and the line of strike is nearly east and west.

The comparative immunity from destruction of Diano, is not difficult to explain. The direction of wave-path was here nearly due north to south. It therefore passed from the deep clays of the piano into the long spur or colline of the town, end on, (see Sketch Nos. 194, 195,) losing a large portion of

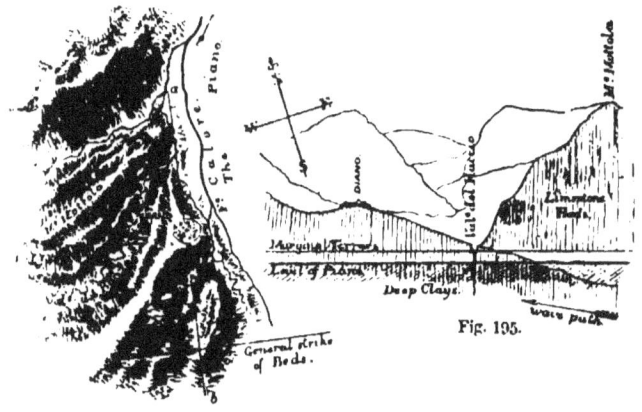

Fig. 194.

its *vis vivâ* at the junction, and a still larger portion, in passing through the great number of nearly vertical beds of limestone, about a mile in total thickness, before reaching the town, in a direction perpendicular almost to their planes, like a bullet shot through the leaves of a thick book.

Again, the shock transmitted southwards through the lengthway of the great flanking chain to the westward, was almost completely cut of from reaching Diano at all, by the Vallone del Raccio to the north and N.W. of the town, upon the S.W. side of which, on the steep slope of Monte Mottola, the effects of the partial extinction of the wave at its surface as "a free or outlying stratum" were visible in considerable falls of projected rock (loose masses chiefly). Nothing of the wave passing along the flanking range reached the town, therefore, but secondary waves of refraction and dispersion, coming up from beneath the town, as the residue of the unextinguished original wave passed southwards.

Few better examples may be found, of the important effects of local condition, as modifying the effects of shock, or of the care necessary to observe and disentangle the phenomena. Of towns situated within three or four miles of each other, one is found almost totally destroyed, the other is scarcely injured. It seems inexplicable at first sight, that both should have been almost equally near, to the same subverting agency from beneath; yet nothing is simpler or more certain when explained, than the conditions which shielded the one, and left the other exposed to destruction.

The protecting circumstances as respects Diano will be understood by comparing Sketch No. 194 with the Section Sketch No. 195, supposing the line $a\,b$ to be that of the wave-path.

St. Arsenio, Torre, and St. Pietro, small places on the west of the Vallone Diano, but north of Diano town, were also more or less protected by similar conditions; cut off from the great flank range, by the little lateral valley of the Aqua del Secchio, and others. They suffered much more than Diano, however, and St. Rufo, on the south flank of the lateral Valley del Torno, still more than either. The wave-path at St. Arsenio and St. Pietro was, from fissures, 142° to 144° W. of north.

Much further away to the westward, (10 to 15 miles west of Diano,) in the heart of the mountains, and in the great extent of rugged country, south of the very high table land of Piano di Salvagnuola, and still more south, in the valleys of the rivers Carmignano, Calore, (another west of the river of the Vallone Diano,) Pietra, and Cilnio, the earth-wave must have been propagated with much violence, but with frequent and rapid changes of direction, and hence rapid loss of *vis vivâ* and speedy extinction. Castelluccia, Ottati, Corbeto, Laurino, and some other towns were greatly damaged. But there are vast tracts of uninhabited mountain and valley about here, and little is known of the shock in these, which I was unable to enter, as they were all under snow of considerable depth.

Still further west, however, after the confluence of the above-named rivers which divide the country for above twenty-five miles in a nearly north and south direction (by compass), though in an irregular line, and previous to the approach of this river to the Bosco Persano, before its junction with the Salaris, the long valley of the western Calore arrested almost completely the violence of the shock, so that between it and the Sea at Pæstum little of it was experienced.

Returning to the Valley of Diano, upon a new piece of the military road not yet used, between Atena and La Sala, was a newly erected culvert, of three semicircular arches of 12 feet span, passing a torrent under the road, the piers, about 8 feet to the springing, all built of good squared ashlar, the arches turned in brick, two bricks thick.

The structure was not overloaded with material and was well put together, and the mortar still green. It did not exhibit a trace of injury. (See Fig. 196.)

Fig. 196.

As I pass southwards, still in the valley, and approach La Sala, the mountain peaks to the rear and above the first low range of the east flank, all show evidence of increased looseness and softness of the limestone rock; its bedding becomes more and more indefinite, its minute structure, more and more like a mass of fine angular compacted fragments, of a rather harder and originally more uniform liassic-looking rock, and its lithological character one of increasing chalky whiteness, with more and more silex intermixed, in the state of a very fine gritty white sand.

The forms of all the mountains behind, the rounded culmination of their summits, the curves of their flanks, with the flowing lines of the ravines, and swelling protuberances of the hill-sides, all alike indicate, a very soft and easily denuded or weathered rock; one of low elasticity and density, and capable of transmitting impulse, much less powerfully, and to a much less distance, than the limestone I had already encountered, and which with its maximum hardness and sonoricity I had found in the flanking peaks of the Valley of the Tanagro, some twenty miles to the north.

Many isolated houses and other buildings about here, founded upon the deep clays of the piano, exhibit by their fissures, an almost completely uniform direction of wave-path north to south, and an angle of emergence so small as to seem almost zero. I observe, however, that wherever such buildings are founded upon the limestone rock, upon the gentle slopes, of the lowest hill sides of the east side, of the Vallone, the wave-path tends a little to the E. of north, *i. e.* it seems to come from the line of the eastern flank range, more or less, but still with the prevailing north to south path; the divergence towards a N. E. to S. W. direction being from 10° to 25°: and the angle of emergence, at once changes from nearly zero, to a pretty large one, but which gradually decreases as I travel south.

Within a mile of La Sala, on the left of the road, stood a square, strong-built house, which was perfectly cardinal, and afforded excellent measurements by fissures and thrown wedges. It gave a subnormal wave-path nearly 171° W. of north, and an angle of emergence of 24°. Another cardinal building to the right of the road just entering La Sala, shows a subabnormal wave-path 155° W. of north, and with nearly the preceding angle of emergence, the wave here again appearing to come from the line of the lateral range to the eastward.

Nearly opposite the town, I observe at a quarter of a mile out in the plain, several large haystacks (Fig. No. 197, Sketches Coll. Roy. Soc.) leaning over at top very much to the southward. They had their longer axes nearly east and west, and all were thrown to the south, without any twist. The people about, said they had been all built plumb, and were so, before the shock, an interesting proof, that a very light body may be overturned by shock, equally with one of great density, the inertia of motion being exactly proportionate to the weight.

LA SALA.—This town, although of Roman, if not of still earlier origin, and showing remains of much antiquity about the old Castello, seen above the town, is nearly all now of modern building; and upon the whole far better built, than any town I have yet seen in the Vallone. It extends for nearly a mile and a quarter, along the slope of the hill-side, the buildings rising above each other, and presenting very generally their greatest length in north and south directions, or parallel to the hill-side, which is nearly continuous, and unbroken by any deep lateral gorges, all along the east side of the Vallone. Not a single stream of any magnitude falls into the Calore from this, but all its feeders from the other or western side. This town is the seat of government of the province, and contains many large official and other structures. All are more or less fissured, but the actually demolished buildings are few. This seems to have arisen less from diminished energy of the shock here, than from the substantial character of the buildings, and from the fact, that almost all of them presented their long dimensions to the line of shock, as may be seen in Photog. No. 198 (Coll. Roy. Soc.) of the town looking from the N.W. The general position of the buildings in plan, as they wound along the hill-side, is along a curve, as in Fig. 199 (Sketch, Coll. Roy. Soc.). There is an elevated little valley, at the back or east of the town, between it and the ridge of the Costa della Madonna, of arid limestone, and slender covering of soil at various points, but no fissures or falls of rock were visible.

I found the Sotto Intendente, Il. Cavalieri Gul°. Calvoso, living with the Signora in a comfortable wooden "barrac" or hut beneath the town; for although the great shock threw down so few buildings here, the alarm of subsequent minor ones, has caused those who could, for the present to desert their permanent stone houses. He accompanied me with his secretary, Il. Caval. Ferdinando Lansalone, through the town, and through his own palazzo (the Casa Officiale), which, though shaken and fissured, was still standing, just as it had been fled from by every living being, on the night of the 16th December; and as it had been locked up ever since, the pictures and many other objects within the house, were lying strewed or thrown about, exactly in the positions in which the shock had left them. The Sotto Intendente gave me on the spot, and in the rooms, a very graphic and intelligent account of his observations as to what had occurred.

There is no record as to the precise time of the shock. The clock at the Casa Communale was thrown down and stopped, but the hour could not be got from it, and its inaccuracy was admitted to be as great as usual.

The house of the Sotto Intendente is founded on the solid limestone rock: it is a long and rather narrow two-story building of large size, stone built, with timber and tiled floors and roof, and well constructed. There are a few small fissures in the walls, indicating a north to south wave-path, emergent $20°$ to $30°$, but the latter evidence is uncertain.

The long axis of the house, has a direction 20° W. of north. The room occupied by the Sotto Intendente with his family on the 16th December, is a nearly square one, on the first floor (*i.e.*, one over the ground floor). They had not gone to rest, and he was first alarmed by a short, sharp rattling, with a jumping vertical movement of about half an inch, of a large white metal chocolatière, that stood upon a marble-topped table at *a*, touching both the south and west walls. At the same instant he heard the "Rombo," which continued during the entire time of the shock. It was not very loud, but very terrible, and "seemed to make the floor and the whole house to tremble "—a hoarse and grating rumble. Before he could have reckoned twenty he thought, the great shock came, a distinct undulation, which several times swayed everything back and forwards, and lifted up and dropped down simultaneously, the horizontal movement having by much the greater range.

As far as he could judge by his own perceptions, the range of horizontal motion did not exceed half a palm (3 or 4 inches). The movements did not instantly cease, after these great oscillations, the total number of which he could not be certain of—he thought they did not exceed four or six—but all was quiet after (as he supposed) about half a minute, when they all rushed out of the house. He never himself, lost his presence of mind; on the contrary, he said that the minutest circumstances of movement, &c., that occurred in the room, from the instant when the chocolatière began to give tongue, seemed to stereotype themselves, upon his observation and memory.

A number of glazed lithographs, in flat wood frames, each hung from a single nail, upon the north and east walls of the room, A and B (Fig. 200). Within a second or two after the chocolatière had begun to jump and make a noise,

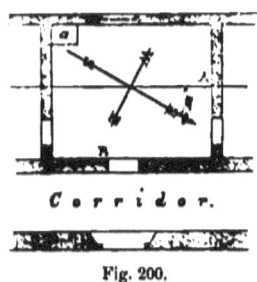

Fig. 200.

the lithographs hanging upon the wall A, began to oscillate slightly in the plane of the wall, or from east to west, and the reverse; and at the same instant, those hanging upon the wall B, began to oscillate slightly, out from and back to that wall, *i. e.*, in the same east to west direction. The great shock now arrived, and the frames upon the wall B, at once began to sway forward and back in the plane of the wall, or in a direction south to north, and the reverse; while those upon the wall A, commenced the movement out from and back to the wall; and for a moment or two he thought they all moved more or less both ways, viz., in the planes and at right angles to the planes, of both the east and west and north and south walls. The motion ended finally, by the prints on the wall B, alone oscillating gently in its plane, with a decreasing motion, for two or three seconds, and finally coming to rest.

Of the lithographs upon the wall A, the Sotto Intendente pointed out to me one, the dimensions of which (they were all quite similar) are given in Fig. 201, and he caused it to vibrate in both ways by his hand, as nearly as he could to the same extent, that he had observed it to have moved at the most violent period of the shock. The chord of the arc of vibration, in plane of the wall (east and west) was about 2 inches, and the semichord of the corresponding vibration, from and back to the wall B, was about 1·25 inch. The chord of vibration of the frames on the wall B in plane of that wall (or north and south), was 7·50 inches,

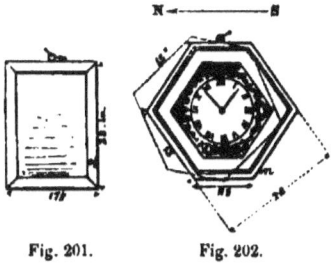

Fig. 201.   Fig. 202.

and the semichord corresponding from and back to this wall A, was about 4 inches.

It was obvious, therefore, that here two wave-paths *almost* coincident in time had crossed each other, at a sharp angle, the one arriving first, being transmitted with more or less of an east and west direction, from the north and south axis of the great range of mountains to the eastward, and having an horizontal amplitude not exceeding 2 inches; the other, which almost instantly followed, having a north to south direction, and an horizontal amplitude of 6 or 7 inches.

Nothing observed, except the chocolatière, gave any approximation to the extent of vertical movement or altitude of the wave which appeared by it, about half an inch at most.

Upon rushing out of the house, Signor Calvoso said, he was struck with the quietude of the night, and the general serenity of the sky for the season. The night was not a dark one, but he had observed nothing himself, of any unusual luminous appearance; of having remarked which, however, numbers of persons had spoken to him since the shock. He however stated that having on the instant, to attend to many official calls, as well as domestic ones, incident to the alarm of all around, he might not have remarked any such phenomena, outside the house.

In his "salone," or drawing-room, a large clumsy pendule in an irregular hexagonal frame, with the dial of about 10 inches diameter, in the centre, like a picture, hung upon the wall parallel with B, in last Figure (*i. e.*, in a plane 20° W. of north), by a single nail and ring at top. A nail driven into the wall at N. (Fig. 202) was in contact with the frame, when the whole hung plumb, and prevented all movement of oscillation towards the S., but it was free to oscillate in the opposite direction, the nail corresponding to N. having been withdrawn from some cause. I found this pendule remaining out of plumb, and thrown to the northward, as shown by the dotted line (Fig. 202), so that the edge of the frame had moved 1·75 inch, from the nail at N. The centre of gravity of the whole, I found by trial, was in the centre of vertical figure, and the weight of the whole was $8\frac{1}{4}$ rotuli = 16·49 lbs., the frame projecting $5\frac{1}{2}$ inches from face of wall. By trial I also found, that the friction of the back against the wall, required a force of $1\frac{1}{2}$ rotuli = 2·91 lbs. to set it in motion from rest. Had it been free of the nail at N., it of course would have vibrated through an arc of 3·50 inches, assuming each semiphase of the wave to have equal velocity, and neglecting the effect of emergence.

From the considerable weight of the pendulum and the large proportion the friction bears to the weight (nearly 1 : 5), it forms by its range of motion an approximate measure of the amplitude of the wave here in an horizontal direction—one too inexact, it is true, to found any calcu-

lation upon, but yet enough to convey a distinct notion of the extent of movement of a powerful earthquake shock.

The actual lateral movement here was probably about 3½ to 4 inches—a range of motion which, made with a velocity as great, as that with which one reaches the ground, on leaping down from a height of 2½ feet, may enable one easily to understand how readily persons are thrown down when exposed to it.

The Secretary Lansalone pointed out to me in his house two brass table lamps, fashioned as in Fig. 203, which stood upon a semicircular table, placed with its diametral side, in contact with a wall running 37° W. of north, the lamps being placed so that the line $x\,y$ was at right angles to the plane of the wall. The centre of gravity of either lamp, is 7 inches above the base, and the centre of oscillation is about 10 inches above the edge of the base, considered as centre of motion. The weight by trial = 1 rotulo 2 unci.

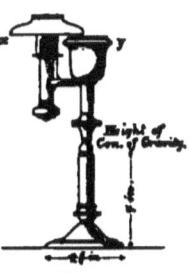

Fig. 203.

At the shock one of these lamps was thrown off the table towards the north and the other towards the south, and lay upon the floor in the positions shown in Photog. No. 205, (Coll. Roy. Soc.), and Fig. 204. The china things upon the same table remained as seen by me, nearly undisturbed: they were low and broad based.

The legs of the table were somewhat elastic, and as they sprung under the shove from the wall, in contact with it, no deduction as to wave velocity can be made from these lamps. Making some allowance for the effects of this elasticity, in giving divergence to the direction of throw, normal to the face of the wall, the position of the lamps indicates a wave-path, about 157° W. of north.

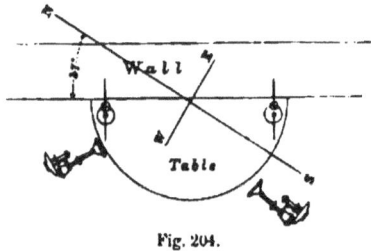

Fig. 204.

There are many pictures in this house in rooms unused, hanging from single nails, which have been caused to swing in the plane of their respective walls. Many remain, just as left by the shock, and all tend to show a general north to south direction, with more or less of movement from a diagonal line approaching east to west. Their friction was too great against the rough walls, to admit of further deduction.

The Sotto Intendente, who, though a keen observer and very intelligent, knows nothing of science or of earthquake speculations, remarks to me that he has observed the buildings situated on the harder limestone, everywhere in his province, much more shaken and injured, than those posited upon the softer chalky stuff found here and further south.

The church of La Sala, like numbers of others in Southern Italy, has been built, not in accordance with ecclesiological notions, but to suit the lie of the ground. Its axial line is 20° W. of north, and it has a campanile, rectangular in plan, external to one flank wall, as in diagram Fig. 206. The upper part of this, previous to the earthquake, carried two bells, hung between the jambs of piers and arches on top. These were overthrown, down to the level of $c\ c$ by

CHURCH OF LA SALA. 343

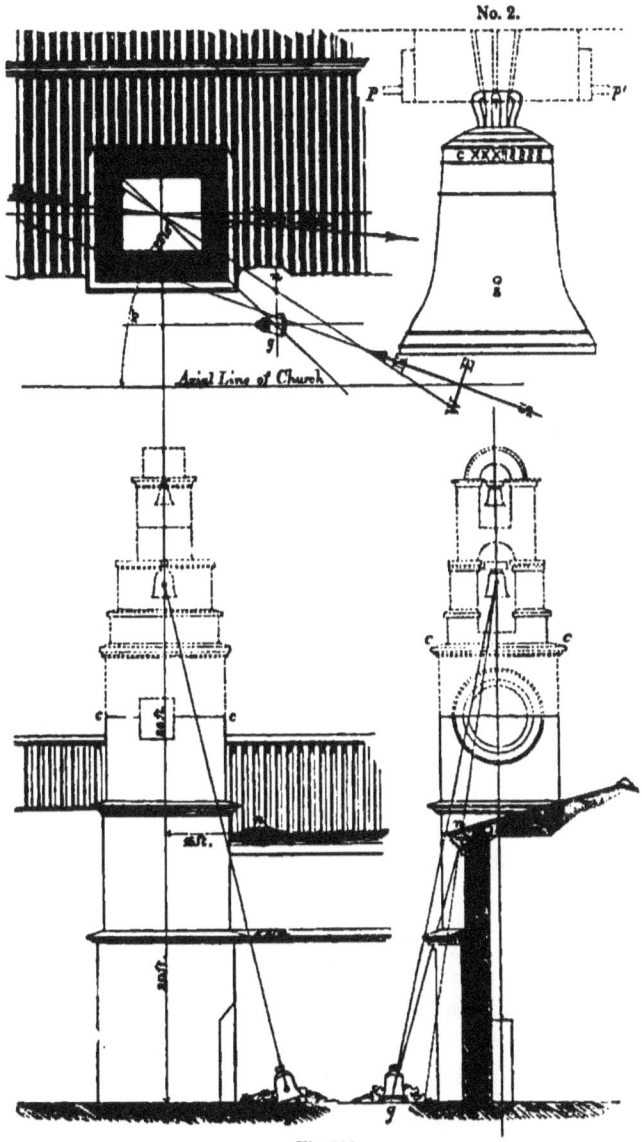

Fig. 206.

the shock, and the tower has since been taken down, to the level of the centre of the mock clock dial. By careful measurements of the standing portion, and by comparison of the taken down pieces, with the sketches and description of the Sotto Intendente, I was able to restore the design of the superstructure, and obtain a close approximation to its original height. Both bells oscillated in a plane, parallel to the axial line of the church. At the shock, whose wave-path passed obliquely through the open arches on the summit of the campanile, the jambs or side piers separated enough from each other, to permit the bells to be released from their pintles, and these were thrown to the ground, in the direction shown in Fig. 206. The larger bell remained unbroken, having fallen amidst some rubbish on the ground; the smaller was destroyed, having descended from a greater height and fallen on hard ground. The piers did not come down at once with the bells, but fell shortly after. The place where the smaller bell struck the ground, could not be ascertained with certainty, but the Padre, the Sotto Intendente, and others, were agreed upon, and pointed out precisely, the spot where the large bell had alighted; and its path of descent was also indicated by the breaches, which it had made in falling, in the eave tiles of the roof, and in those of a string course of the external flank wall of the church, near the campanile. The bell itself had been removed into the church (but the fallen rubbish was still on the ground): it is very antique and remarkable for its form, and is correctly sketched in Fig. 206, No. 2. It bears an inscription in Lombardic (?) characters, and a date which I presume to be 1324 or 1336. The Padre, who is accustomed to judge of the weight of bells, estimates it at 7 cantari = 1225 lbs., which agrees with my calculations from volume.

The appearance of the church and tower is shown in Photog. No. 207 (Coll. Roy. Soc.), as seen from the N.W. Nothing can be inferred from the fall of this bell as to the direction of the wave-path (which we have already got from other data), inasmuch as it was obvious that a line drawn from the centre of the tower to the centre of gravity of the bell, at its place upon the ground at g, would be far from representing truly the plane in which its trajectory of descent had been made. It had first descended to n, and coming in contact with the eave tiles, had then taken a new course, and been thrown directly outwards or westward from the wall, in its further descent.

It was also observable from the scratches, &c., on the bell, that the eastern pintle had given way first, so that the full force of the shock had not acted in projecting it. The shock acting at the centre of gravity s (No. 2), and the resistance of the last held pintle p, had caused the bell to rotate slightly before falling, and given it along with its glancing off the eave tiles, a much more westerly direction of descent, than was due to the direction of wave-path alone. Both these extraneous forces also had reduced the horizontal distance to which it would have been thrown, had it freely pitched to the ground in its original path—both pintles being freed together.

Taking the vertical height of fall to be that from the centre of gravity of the bell as it hung, to the level of the eave tiles, and the breach therein as marking the horizontal distance thrown, and applying the equation—

$$\operatorname{Tan} e = \frac{2H + \sqrt{4H(H+b) - a^2}}{a}$$

we have $b = 26$ feet, and $a = 16$ feet as measured. We must estimate the deficiency of a from the above circumstances, and, adding $\frac{1}{16}$, assume $a = 17$ feet.

Taking the velocity of the wave as we found it at Polla = 13 feet per second,

$$H = \frac{v^2}{2g} = 2\cdot 62, \text{ the height due to V,}$$

and solving for e — we have

$$\text{Tan } e = \frac{5\cdot 24 + \sqrt{299\cdot 9 - 289}}{17} = 0\cdot 513$$

$$\text{and } e = 27°\ 10'$$

which is the angle of emergence, as given by the projection of this bell. This, owing to the disturbing conditions, can only be considered as approximatively true, to within ± or 3°. The result, however, is corroborated by the same angle having been obtained otherwise, as already given: better data were not procurable at La Sala.

The base of the church tower may be viewed as about the mean level of La Sala. The barom. reads 27·86 in. Thermo. 54° at $11^h\ 0'$. A.M. Naples time, 15th February, which reduced gives the altitude of the spot = 1768·30 feet above the sea, and about 1000 feet above the piano at its northern extremity.

The cause of the disturbances in the general direction of wave-path observable at various points since leaving the north end of the valley, began now to be apparent, on examining with the maps, the relations of the lateral chains at either side, to the main direction of shock hereabouts.

Referring to Fig. 208, in which is sketched the general form of the piano, and the prevailing lines of ridge, or summits of the mountain ranges around it are marked by dark lines; the main direction of the wave-path from Campostrina southwards is indicated by the large arrows $w, w,$ and is that given generally by everything in the piano, south of Polla, to Padula. But on the east side of the valley, going southward, it is constantly disturbed by the intermixture of another wave-path oblique to the former, and with increasing obliquity the further we go south; this intersecting angle, being small at Atena, greater at La Sala, and greater still at Padula, and the direction of wave transit of this secondary shock being always from N. E. to S. W., or *from* the lateral chain to the east of the valley.

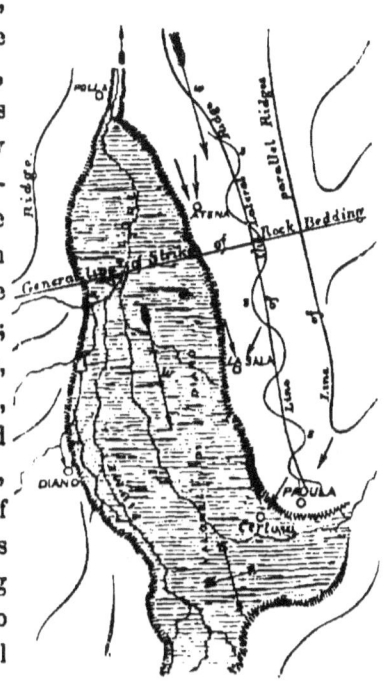

Fig. 208.

It is obvious that the impulse given at the north end of the valley in a north to south direction, simultaneously to the deep clays of the piano, and to the eastern mountain range at its north end, was transmitted comparatively undisturbed as to direction through the former; but, in passing along the lateral ridge of limestone mountain across all the nearly vertical beds, transverse vibration was produced along the whole line of mountain crest, which oscillated transversely, like a thin plate or musical string, struck at one point, giving rise to a secondary wave whose path was that of the wavy line *s, s, s*; the transverse or east and west excursions becoming greater the further the originating wave travelled south, and being both exaggerated and disturbed in direction, by the many curves and sinuosities in the mountain axis not shown in the sketch. Every point situated upon the limestone rock, therefore, at the east side of the valley was exposed to two shocks; the primary in the direction nearly north to south, and the secondary (from the transversal vibration of the mountain chain) more or less oblique to that; and these arrived, not quite simultaneously, but with an interval of time, as well as an angle of intersection, greater as the point lay further south. Such transversal vibration, of the almost continuous though sinuous ridge to the east of the valley, was of course not confined to the horizontal. The chocolatière of Sig. Calvoso, at La Sala, gave evidence of the same in the vertical, and also shows that the originating wave traversed faster through the limestone rock, notwithstanding its loss of *vis vivâ* in penetrating across from bed to bed of east and west stratification for so many miles, than through the deep clay of the Vallone. An examination of the sketch will also indicate the reason why the towns at the west side of the Vallone received so much less damage than those on the east side. All the feeders of the Calore fall in upon the west side, not a single one upon the east; the west lateral range is therefore cut across by lateral valleys, down nearly to the level of the piano, while at the opposite side the range is continuous. The wave therefore was either diverted westward, or nearly extinguished at the western side of the valley, before it had reached far to the south, while at the eastern side it was carried on by the continuous lateral chain until changed in direction, and greatly diminished (by the free stratum) where the continuity of this range is broken and terminates, near Padula.

About a mile south of La Sala, near the military road, and standing at the edge of the piano upon clays of a few feet in depth, overlying the limestone lowest roots of the hills where they dip beneath the plain, stood the Chiesa della Trinita; the short tower, the roof and west end of which have all fallen, and the walls are fissured severely. It is seen in Photog. No. 209 (Coll. Roy. Soc.) looking westward. The tiled roof had fallen within the walls, leaving the tiled eaves still upon the tops of the side walls: its debris had been removed, but I was informed that the larger portion was found packed against the north flank wall upon the floor. The west end was thrown to the S.W. The east end had been kept up by cross walls external to it. It presented fissures, best seen at the east side of the wall (not shown in Photog.) which showed an emergent angle of 23°. The wave-path could only be approximated, but was a few degrees W. of north to south.

The Ponte Silla, an ancient Roman bridge over the Calore about a mile and a quarter south-west of this, founded probably on piles, with its heavy piers deep in the clays of the valley, has suffered no apparent injury by the shock.

CHAPTER X.

PADULA AND ITS NEIGHBOURHOOD—THE PALAZZO ROMANI
AND ITS GARDENS.

PADULA, a large town, said to be of Lucanian origin, but with no traces of extreme antiquity, and generally well built, stands upon a pretty steep eminence of solid limestone, at the southern extremity of the first great break in the lateral chain to the east of the Vallone. The range rises high above to the N.N.E.; and eastward of the same there is a narrow lateral valley in a direction about N.E., and exactly at the tongue, between it and the great Vallone stands Padula. The Vallone here narrows very rapidly, and trends off to the westward at about a mile or two south; and still further south at Buona Bitacola on the west, and Montesano on the east, it becomes quite hemmed in by lofty mountains. Violent contrary motions must therefore have been produced by the breaking up of the great wave of shock in the deep alluvium, and its in-baying and shelving up upon the irregular masses of limestone mountains that divide the many divergent valleys, and gorges, and project between them. Directly south of Padula, and of the great Cistercian Monastery of St. Lorenzo, about a mile west of the town, the clays of the piano give place to a rolling and broken surface of clay and gravel, overlying limestone rock at a greater or less depth, and extending for more than four square miles; above which to the east rise the summits of Monte St. Elia, and Monte della Vajana, which recommence the broken chain of the E. lateral range. Padula therefore, has got, soft, cretaceous, and sandy, white limestone beds with east and west strike, and nearly vertical stratification to the N., similar but harder limestone of unknown stratification to the south, and the deep clays of the piano to the west of it. The rock upon which the town stands is of the same quality and stratification as the range to the north, and the colline is joined on to the great range by a shoulder at rather a lower level. The western and southern slopes of the town are very steep, averaging probably 30° from the horizon towards the plain; the buildings generally, either founded on the bare rock, or upon a thin stratum of diluvial matter, which increases in depth as we descend towards the plain.

The buildings on these sides are the oldest and worst, and have suffered the most, having been exposed to the severest brunt of the shock, and been the least able to bear it, and in the most favourable position heaped above each other to produce mutual destruction in falling. The counterscarp of the town, or that on the east and N.E. towards the narrow valley and gorge behind it, is less steep close to the town, but within a quarter of a mile of both sides of the gorge becomes precipitous; the west side being bare limestone rock nearly to the bottom, which is not much above the level of the piano (at the town), and the opposite, or east side, covered with steep banks of calcareous clays; limestone, angular gravel, and sand, getting very deep and heavy as they descend to the bottom, where they are being cut into and carried away by the torrent and rivers round the town to the east and south, and passing the Certosa de S. Lorenzo farther into the Calore. This torrent issues principally from copious springs in the rock and coming from under the clays, about a mile and a half up the gorge to the north; and these are now, and have been ever since the earthquake, highly turbid and discoloured by the reddish earth, and are said by the Syndic and Subjudice of Padula, who visited the place with me, to be largely increased in volume of water, since that event.

The Photog. No. 210, is taken in this valley, looking back at the town toward the S.W. On the steep counterscarp to the right of this view I found a large fissure in the solid clay covering, extending some 300 yards in length at about 270 feet above the bottom of the gorge. The direction was due E. and W. by compass, and its S. slip was from 1 to 7 inches below the level of the opposite one. The fissure is a flowing curve similar in horizontal plan to the contour of the hill side, and its line of direction is just that most favourable, to a throw off and slip of the clay masses, upon the sublying rock, by the jog of the earthquake, which is unquestionably the nature of its formation. Upon the opposite side, and about half a mile up the valley to the north, where the east slope has become much steeper, I observe with the telescope that huge masses of clay and gravel that had stood above the torrent as nearly vertical banks, of from 50 to 120 feet in height over the water, have in several places been shaken down, and fallen in great masses, damming the torrent into large and very deep, discoloured pools, from some of which the dams have already given way by little debacles, while in several of the others, the water is escaping beneath and through the clay, and carrying volumes of fluid mud and sand away

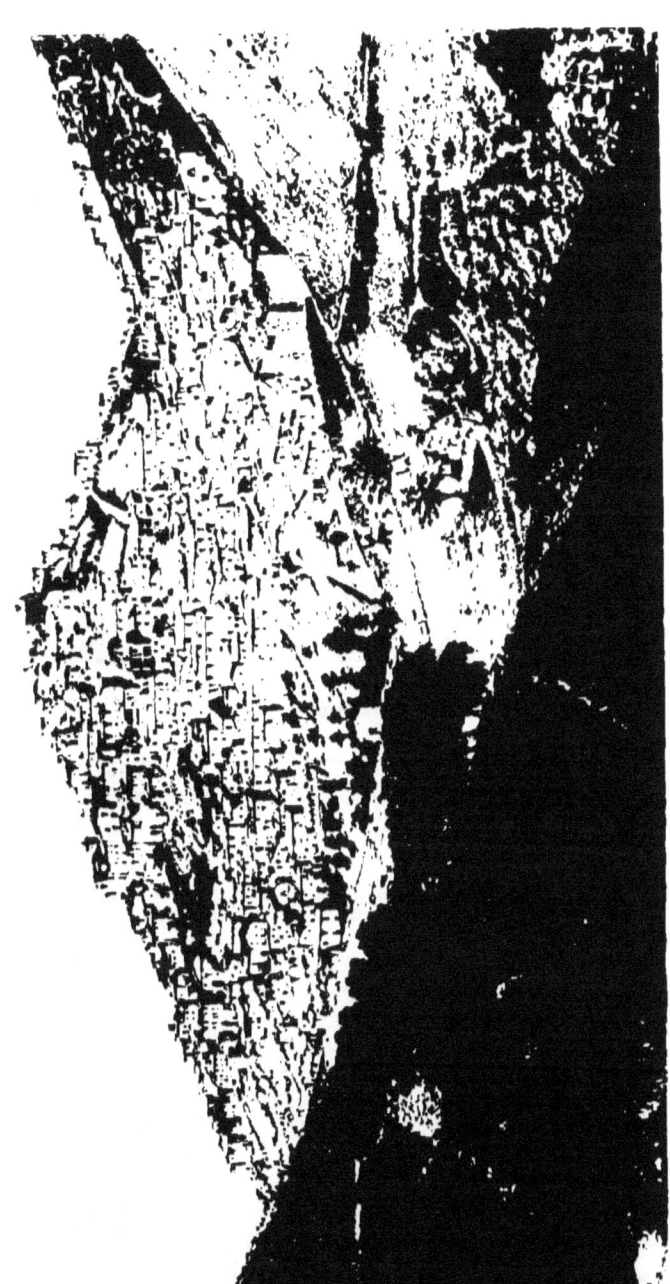

VALLEY TO THE EAST OF LA SALA. VAL. DI DIANO.

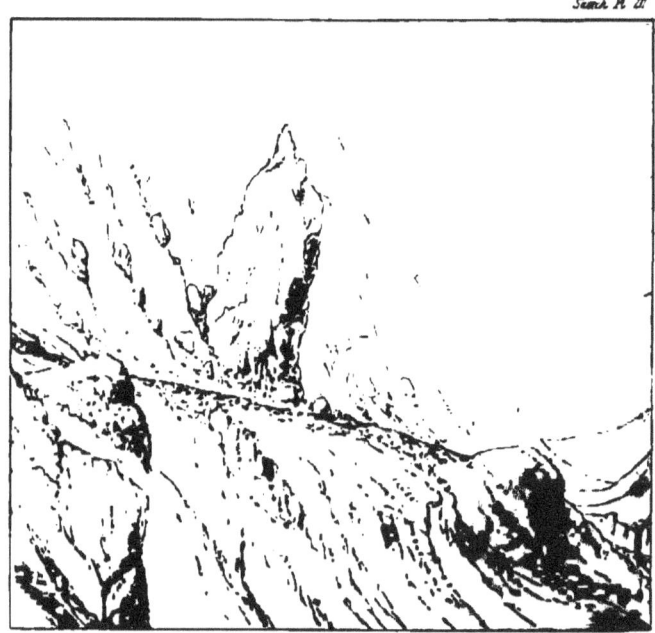

ROCK AIGUILLE near PADULA.

ROCK AIGUILLE near PADULA, fractured from its base.

with it: a very few weeks with heavy rain, will carry tens of thousands of tons of clay and gravel down into the Calore.

Higher up the gorge than the Photog. view extends, and upon the east flank of the gorge, this slope becomes very precipitous, and consists of bare cretaceous limestone, standing up in aiguilles and isolated weathered masses, some of great magnitude, and here (as well as from some points on the rocky counterscarps of the town colline), ponderous falls of rock have taken place. The massive fragments with newly-broken and glittering white surfaces encumbering the slope, or, after having traced their descending paths in lines of torn rock and furrowed detritus, block up the bed of the torrent below, which brawls between the immense fragments and beneath them. Most of them have been detached from the outcropped ends of the ill-discernible vertical strata, separating at joints, &c., and are mere cases of loss of equilibrium by the shock; but one most remarkable case I observed, in which an enormous mass of solid rock that had stood up as a sort of blunt aiguille from the steep face of the slope, not quite vertical, but rather overhanging, (about 15° by the eye,) to the downward side, had been broken clean off at its base, and again breaking into three massive pieces, had slid down the rocky slope, and now occupied a place about 150 feet below, having crossed and wholly torn away the mule-path in their progress. The appearance of these masses, as they must have been prior to their fall, may be gathered from Sketch No. 211, and the way in which the fragments lay from Sketch No. 212.

The aiguille, by measurements of the three principal fragments, must have stood about 70 feet high above its base when entire, and the three masses, when roughly cubed, contained fully 35,000 cubic feet, and weighed about 2,500 tons. The surface of fracture, of which about ⅔ths was formed of planes of separation (transverse joints) in the rock, and the remainder through the solid stone, was an irregular circular figure, and about 20 ft. × 18 ft., the plane of fracture generally being inclined about 45° to the horizon.

The rock was very soft, arenaceous, and chalky limestone, perfectly white, and a hand specimen, when fresh fractured, could, like sugar, be rubbed away against another piece; so that the cohesive energy was not great, and the inertia of so great and high a mass was quite sufficient, at a very moderate velocity, to bring it over. It however has a peculiar interest, as the first example I have seen of actual rock fracture by the direct operation of the shock.

Neither the position of the centre of gravity, nor of that of resistance of the fractured base could be fixed with sufficient accuracy to enable any calculation to be based upon the fracture and fall of the mass, that would give a trustworthy measure of velocity or of direction.

On examining the buildings in the town on the south and south-west slopes, where the damage done was greatest, I found evidence by fissures and projected wedges of masonry, giving a direction of wave-path varying between the limits of 155° E. of north to 167° E. of north, and angles of emergence (all, however, proving emergence from the north and N. W.) varying between 20° and 25°. Some of the fractures from which the latter elements are taken are visible in the Photog. No. 210, though not so in its lithographic reproduction.

Many circumstances indicated the transit of two shocks crossing obliquely, such as the twisting of objects on their bases, and the almost universal occurrence of fissures in *all* four walls of rectangular buildings, whether cardinal or ordinal.

At the very top of the town, founded upon the bare limestone rock, which here protrudes to the surface everywhere, I observed an old and partly ruinous mansion, the Palazzo Romani, which, as well as its deserted gardens, afforded me some valuable data.

The palazzo is a large, nearly square, three-story building, very nearly cardinal. The walls are well built of rubble. It stands on level ground. It is fissured in all four external walls: all the fissures are long and threadlike, and are scarcely visible in the Photogs. No. 213 and No. 214 (Coll. Roy. Soc.), the former showing the S.W. quoin, the latter the opposite, or N. E. one. From previous indications of double shock here, I dare not conclude anything as to horizontal component of wave-path, from the fissures in adjacent walls; but those in the flank walls gave excellent evidence as to angle of emergence, which proved to be 25° to 25° 30′ from north. The fissures in the adjacent walls were much more nearly vertical (8° to 10° inclination), and hence appeared to have been produced by the secondary shock, which shook the whole colline of the town to its base, and therefore by an oscillation more nearly in an horizontal path. I could only gain access to part of the interior; it was empty and disused, and offered little to record, probably.

In front, and to the N.W. of the palazzo, is a little columnar monument—Il Croce Romani, of the ordinary character; a Roman Doric column of 9 inches diameter of shaft and 9 feet high, supporting a white marble ball and cross, and placed upon a square pedestal and plinth block, elevated on a few courses of rude masonry as in Photog. No. 215 (Coll. Roy. Soc.), and Figs. 1, 2, and 3, Diagram No. 216. The ball and cross are secured to the top of the column, only by an iron dowal, and are loose; the plane of

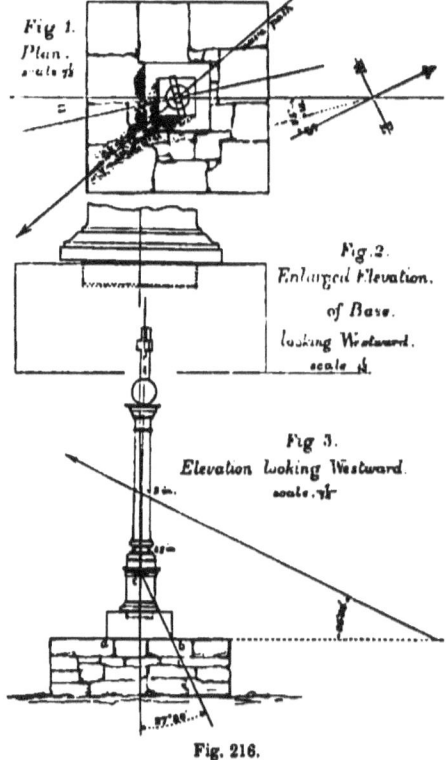

Fig. 216.

the cross has been long twisted out of square, I was informed.

No injury has been sustained by this little structure, except that the pedestal has been thrown over, being lifted slightly at the south side by ¼ of an inch vertically, as in Fig. 2, and jammed quite close to the plinth block at the opposite or north side. The shaft, pedestal, and plinth are socketed into each other, as shown, without cement.

A vertical plane passing through the axis of the column, &c., in a northerly and southerly direction ranges 25° E. of north. The lower edge of the pedestal is lifted more at the western than at the eastern corner of the south side, so as to infer a wave-path about 10° or 12° E. of north.

The shaft and pedestal, &c., are all of hard Apennine limestone, and the cavetto moulding above the base is 8½ inches diameter only : an inference may be drawn from this as to the maximum possible velocity of shock, emergent at the angle we have found, viz., 25°, that would have left this little column unbroken at the neck formed by the cavetto. The horizontal velocity for fracture only is given by the Equation XXV.

$$V = \tfrac{8}{3} g \times \frac{LD}{a^2};$$

where D, the neck of the cavetto = 0·708 feet, a, the height of the column adding in the ball and iron, to the cylinder = 6·5 feet, and L = the length of the modulus of cohesion for the material. This will be = 225 feet, if we take the weight of a cubic foot = 160 lbs., and its cohesion at 500 lbs. per square inch (which is supported by Hodgkinson's experimental determination for marble, 551 lbs.). Then

$$V = 20·12 \times \frac{225 \times ·708}{42·25} = 75·961 \text{ feet per second},$$

and

$V \sec e = 75·961 \times 1·103 = V = 83·785$ feet per second ;

the velocity at the emergence found, that would have *just fractured* the column at the cavetto.

It is, therefore, obvious that the velocity of the shock was greatly below that that would have been necessary to fracture the column by its own inertia of motion, even if the cohesion which I have assumed for the limestone of which it was formed be much above the truth, as is probably the case. The calculation, however, removes the wonder with which the unaided senses regard so slender a stalk of stone found whole and uninjured, in the midst of bowing roofs and massive walls rent or overthrown.

In what had once been the garden of the palazzo stood a portion of a shaft of an old column—one of six—that had once formed some sort of garden edifice. It had been broken off and overthrown, and afforded a precious admeasurement both for direction of wave-path and velocity of shock.

The columns had all been originally formed of an artificial "beton" or concrete, of broken limestone and brick, and lime mortar, rudely formed enough, without any base moulding but a fillet, and were 17 inches diameter. They stood upon the top course of a continuous base of limestone, running round a large circular raised platform of earth. They had all been broken down and the shafts had disappeared to within a few inches of their bases years since, except this one portion of a shaft, which stood before the earthquake 4 feet 8 inches above its base. For 6 inches in height the shaft was a united block, and fast to the stone base; at this height an old fracture had existed, and a portion of shaft, 2 feet 10 inches in length, had stood upon the lower block, having been replaced upon it, and a little fine mortar interposed in the old fracture. In toppling over, at the north side of the column, the arris was broken out, and the diameter in N. and S. direction reduced to $15\frac{1}{2}$ inches, as shown in Figs. 5 and 6, Diagram No. 221. Upon the

top of this block, had stood another piece of the shaft
1 foot 4 inches high, cemented to it, in like manner at an old
fracture. When examined by me, the two upper blocks had

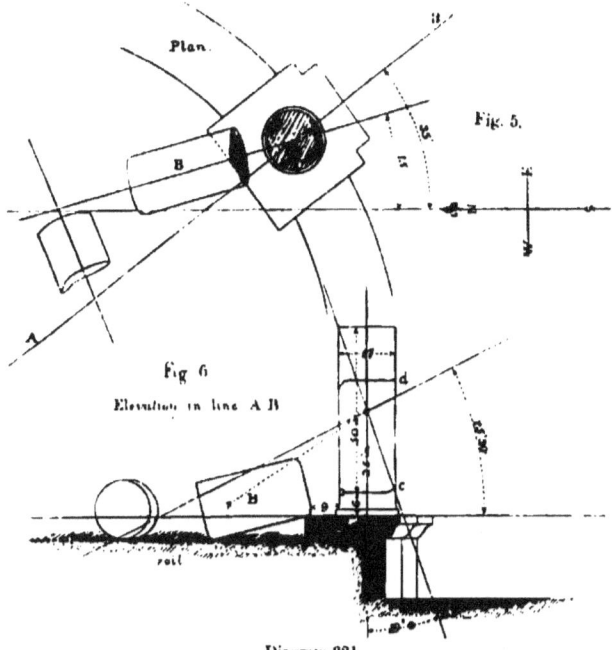

Diagram 221.

been overthrown by the shock, and lay as in Figs. 5 and 6,
Diagram 221, and as shown in Photog. No. 217, having never
been moved or meddled with, the garden being an enclosure,
since the earthquake. The lower and longer block was, at its
lower end, within 9 inches of the stump of the shaft. The
top block had been thrown over along with the one below
it, had separated on striking the ground, at the top part, as
the position of the other piece and the lower level of the
soil than the stone base (Fig. 6) proved, and had slewed
round and rolled a little off. The longer piece remained

held by its impression in the soft soil precisely in the spot in which it had fallen.

The mortar joint at c, Fig. 6, had obviously possessed scarcely any bond whatever, and the only resistance to fall by the shock was, therefore, the stability of the superimposed two blocks upon the base or joint c. The form of the surface of separation, showed that the blocks had overturned precisely in a vertical plane, passing through the axis of the longer block as found prostrate, which corroborated the assurance of the Syndic and others present, that they had not been moved by any one since their fall. The direction of the axis of the fallen block, B, was exactly 15° W. of N., and such was the wave-path that overthrew it, the movement having been from N. to S.

We may view the velocity that overthrew the shaft, separately from that which projected it from its base to the horizontal distance given, and compounding these two horizontal velocities, and resolving in the direction of the wave-path, obtain the total velocity impressed upon it, and hence that of the wave itself.

And first we obtain the horizontal velocity for *overturning* only from the equation

$$V^2 = \frac{15\,b^2 + 16\,a^2}{12\,a^2} \times g \sqrt{a^2 + b^2}\,(1 - \cos \theta).$$

Here $a = 50$ inches, $b = 17$ inches, $\theta = 19°$; and solving for V we have

$$V^2 = \frac{44335}{30000} \times 2\cdot 683\,\sqrt{2789}\,(\cdot 0545)$$

$$= 1\cdot 478 \times 2\cdot 683 \times 52\cdot 81 \times \cdot 0545$$

$V^2 = 11\cdot 42$ and $\therefore V = 3\cdot 38$ feet per second, which is the *horizontal* velocity necessary to *overturn* the shaft only.

But it was also *projected*, so that a point taken at the lower arris of the overthrown shaft had moved horizontally a distance of 9 inches from the corresponding point of the base above r (Fig. 6, Diagram No. 221), while the same point had descended vertically 6 inches from the arris of the base, as already described. Calling the horizontal ordinate A, and the vertical one B, we obtain the *horizontal* velocity of *projection only* from the equation

$$V^2 = \frac{g}{2} \times \frac{a^2}{b - a \tan e}.$$

Here $a = 0\cdot75$ feet, $b = 0\cdot50$ feet, $e = 25°\cdot30'$

$$V^2 = \frac{32\cdot2}{2} \times \frac{0\cdot563}{0\cdot500 - 0\cdot750 \times 0\cdot4769}$$

$$= 16\cdot1 \times \frac{0\cdot563}{0\cdot142}$$

$V^2 = 63\cdot756$ and $V = 7\cdot98$ feet per second;

adding the two horizontal velocities thus found, we obtain the *total horizontal* velocity impressed upon the shaft; or

$$V = 0\cdot38 + 7\cdot98 = 11\cdot36 \text{ feet per second.}$$

But this must be resolved to the direction of the wave-path, whose angle of emergence here we found to be

$$e = 25°\cdot30'.$$

The velocity in this direction therefore is

$$V \sec e = 11\cdot36 \times \sec e,$$

or finally

$$V = 11\cdot36 \times 1\cdot108 = 12\cdot586 \text{ feet per second,}$$

feet per second, which agrees very closely with previous determinations from other objects and at other stations.

The velocity thus obtained may be subject to two slight corrections, for which, however, the data are not obtainable. 1st. If the mortar at the old fracture of the base had *any* adhesion at all, (I believe it had none,) the velocity due to *fracturing* it, should be added.

2nd. The horizontal ordinate a in the last equation is probably a little too great as measured, inasmuch as the fallen piece of the shaft must, by its own elasticity and that of the ground, have *slid forward* some fraction of an inch after it had struck the latter, which would make the velocity obtained a little too great.

These two errors, if they exist, tend to correct each other; and neither could affect the result to the extent of unity in the first decimal place. The result may therefore be relied upon.

In the same garden two vases of limestone, rudely hollowed out, had stood upon the opposite corners of the low parapet walls, that once confined the soil of beds, and separated them from broad walks between, now overgrown with dense turf and weeds, moss, twigs, &c.

These vases were both projected off their feet or lower portions, at a joint at e, Figs. 1 and 2, Diagram No. 222, and thrown to the ground, the lower portions remaining still upon the parapets, as seen in Photog. No. 220 (Coll. Roy. Soc.). One of the lower portions, (A, Fig. 1,) I found square to the faces of the parapets, and obviously unmoved; the other, B, had been twisted round 14°, so that two of its sides were in a line exactly north and south 14° W.

The lower portions at the joint e (Figs. 1 and 3) were quite

IN THE GARDEN.

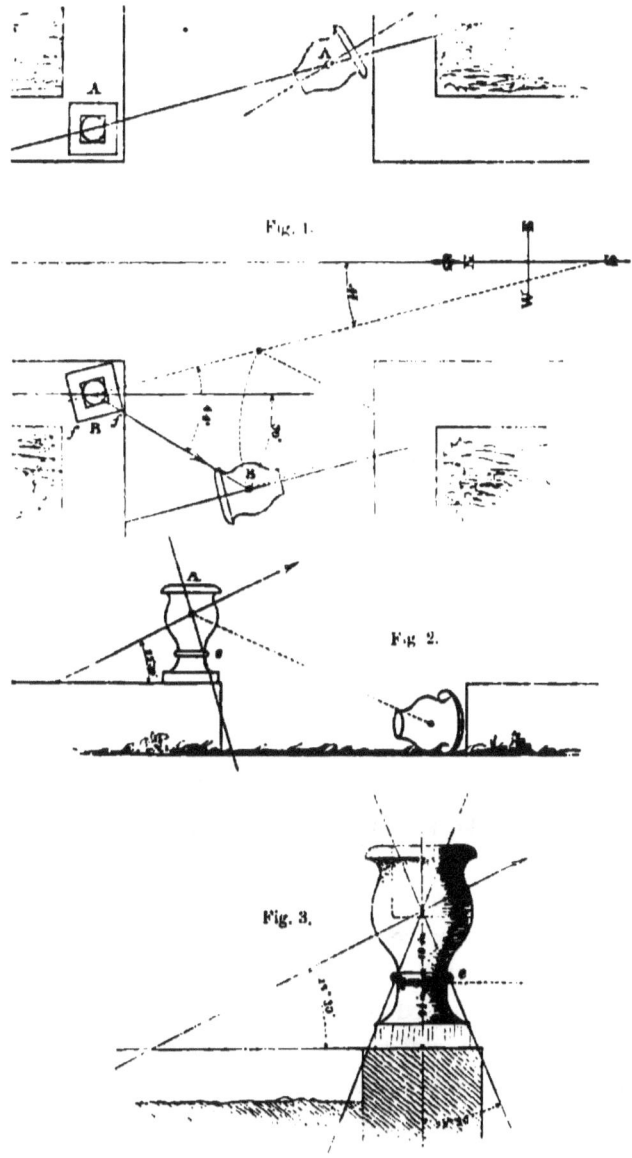

Fig. 1.

Fig. 2.

Fig. 3.

Diagram 222.

level and smooth, and the arrises perfect; and the upper and lower parts had been merely laid together, without any cementing material.

All the parts, like those of the column, &c. preceding, had remained untouched since their fall by the earthquake.

The centre of gravity of the upper part of each vase, (that projected off,) I found by trial was at 10 inches above the joint e, and the weight by trial, was 94 rotuli = 183 lbs. avoir. nearly.

The vase A had been thrown so, that its centre of gravity had been displaced 6·50 feet in the horizontal, and 3·60 feet in the vertical direction. We get the velocity of projection from the equation

$$V^2 = \frac{a^2 g}{2 \cos^2 e \, (b + a \tan e)}$$

in which $a = 6·50$, $b = 3·60$, $e = 25°·30'$, as before. Then

$$V^2 = \frac{42·25 \times 32·2}{1·629 \times (10·1 \times 0·4769)} = \frac{1360·50}{7·85}$$

$$V^2 = 173 \therefore V = 13·152 \text{ feet per second,}$$

which only differs from the velocity given by the fallen column, by 0·566 foot, or a little more than 6 inches per second.

As respects horizontal direction, the vase A had been thrown in precisely the same path as the column preceding, viz., 15° W. of north towards the south; and its axis of symmetry was twisted as it lay, about 12° in a horizontal plane, from the vertical plane of projection, as seen in diagram, and it was obvious, from examination of the truf beneath, that it had not rolled after it struck the ground.

The vase B, however, lay (as in Diagram) so, that its line of direction with its own base, was 30° E. of north, the horizontal range of throw being only = 4·75 feet. There was evidence of this vase having rolled, from the point where it first alighted, westward, but the precise point at which it first struck the ground was not decisive. It appears probable that it was first, along with its base, tilted up upon the edge ff by the transverse shock through the limestone, and that before it had returned to its former position, the principal or direct shock, in the direction 15° W. of north to south, caught it and threw the vase off the base; but part of the effort of projection was in that case destroyed by the fall back of vase and base, and hence the horizontal distance of projection is less than that of the other vase, and less than that due to the velocity of shock; and the direction of projection became one in some intermediate azimuth between the two wave-paths. We might assume that the vase, B, was projected by one shock, and that, A, by the other; in which case we could infer that the angle of horizontal intersection of the wave-paths had been = 45°, but that would render no account of the twisting of the base of B. We must, therefore conclude, that the vase B was acted on by both shocks, and as it had rolled more or less, the angle of intersection would be less. The supposition made, also gives a solution for the fact otherwise hard to account for, that the part of this vase that rested on its base was towards the south as it lay, while the corresponding part of A faced the north. The latter was a *clean throw*, in the former the vase was *tilted* a little towards the N.E. in the return to its place, of its partly-upturned base, immediately previous to its throw, and the rotation thus commenced, turned it over in a vertical plane during its free descent from its base to the ground.

It will be seen, on looking at Fig. 3 (Diagram No. 222), that the vase might have been thus tilted upon its own base at the joint *e*, to the extent of nearly 22° 30′, before it would have lost the power of recovering its position, and that when tilted through part of this angle, the effect of the throw of the second shock delivered to it through the base, would tend to increase the rotation commenced, as well as to diminish the extent of horizontal range, or velocity impressed.

The general relation of all these objects as to azimuth, is shown in Fig. 223.

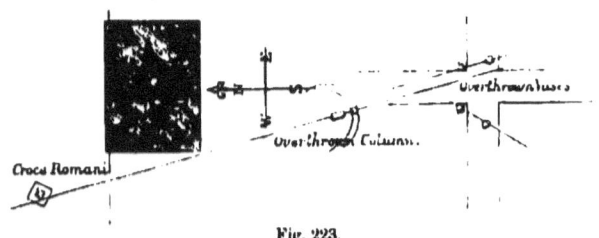

Fig. 223.

I found here also, as at Polla, an example of that singular circumstance, the keystone or block, of a cut stone semicircular arch over a doorway, which had *worked up* in place of coming down (by the movements of the earthquake) between the remaining arch stones. This case, of which the (original) Sketch No. 224 is a true representation, (though very imperfectly given by the woodcut,) is obviously due to the rocking to and fro of the whole wall in the plane of the arch, the motion being several times repeated, and hence the alternate partial freeing and gripping of the keystone, at *a* and *c*, *b* and *d*, between the rocking voussoirs, which, moving on the lowest point of the jambs, or at the springing level, are thus at each oscillation, relatively higher and lower than each other, at *d* and at *a*. The keystone at each alternation, if already tolerably free above, by the fractures of the wall, produced at the first moment

Fig. 224.

of shock is thus moved up more or less, and finally remains at the height gained.

In this instance, the keyblock had been thus lifted up, 1¼ inch above its former place. It is an excellent example, of one of the many cases, in which a misinterpretation of the phenomena, or misconception of the forces, &c., produces false notions as to the actual movement, of the shock productive of them. My attention was drawn to these arches, here and at Polla, by intelligent men (Syndici and Judici, &c.) as affording proof positive, of a sudden drop down, or jump up vertically, of the earth and the buildings upon it.

CHAPTER XI.

PADULA AND ITS NEIGHBOURHOOD.

BELOW the town, and nearly upon the level of the piano, and founded upon the deep clays, stands the monastery of St. Francisco. Its greatest length stood transverse to the general wave-path, and it has suffered much. Its walls present fractures complicated by the double shock; but those whose planes approached a north and south direction, gave good measures of angle of emergence, which appeared a good deal less, down on the deep clays, than up on the rocky eminence of the town.

The average of the measurements taken from fissures and fractures, some of which were in the portions of the monastery seen in Photogs. No. 223 and No. 224 (Coll. Roy. Soc.), gave an emergence of 18° to 20° from the north, or $5\frac{1}{2}°$ to $7\frac{1}{2}°$ less in the clays than in the limestone rock.

The Syndic of Padula, who accompanied me over the whole place, was of opinion that the great shock came from the northward, but that it was also "vorticoso," or at least in various directions transverse to the main one, and so close together in time, that it was impossible to regard the earthquake (here) as other than a prolonged and irregular succession of oscillations, lasting several seconds, he could not say how many. The second distinct shock was about an hour after that; they had no means of telling the exact time of the occurrence. He heard the sound, he thought, about the same instant that he perceived the first movement. It was "a deep rolling murmur," and lasted as long as the movement. In these statements I found the Judice, and three or four of the better class of inhabitants who accompanied us coincided.

From the town and its neighbourhood I proceeded about a mile and a quarter to the magnificent monastery, the Certosa de St. Lorenzo. Where, then as well as on my return from the country further south and west, I was lodged with a graceful hospitality deserving of record. Within those quiet walls I remained two days for the purpose of writing up my journal from the pencil note-books, whose legibility was not to be trusted, having been written, for four or five days past, under almost continuous rain; and also to examine carefully the many instructive damages which the vast building had sustained. In both of these, I was inconvenienced by severe swelling and acute rheumatic inflammation of the backs of my hands, produced by their constant exposure to the wet, &c.

This noble old monastery (whose size and architectural grandeur rendered it worthy of lodging royalty, before it had been despoiled and defaced, in the French occupation under Murat) is built wholly of the best and hardest quality of the white limestone of the higher adjacent mountains, and founded altogether upon the deep clays and gravels of the piano. An eye-sketch plan, of that portion of the whole mass of buildings which I examined seismically, is given in

Diagrams Nos. 238 and 240. Its architecture will be gathered from the several Photogs., &c. It is shattered and shaken to its foundations in every portion, by the violent effort of the earthquake, and presents characteristics of much more formidable dislocation, than the town of Padula did.

All its walls, its vaulted church, and refectory, and very many other grand roofs, the noble groining of its cloisters, and the painted and richly stuccoed ceilings of its library, and of many a royal chamber, are split, fissured, and falling. The light and the rain now find their way through acres of shattered tiling. Innumerable chimneys, obelisks, parapets, vases, bassi-relievi, statues, are thrown down, disfigured or destroyed. Even the internal framing of the heavy-timbered roofs, has been in several places crushed, by the fall of heavy masses from above. Nearly all the superb columnar arcades around its cloistered courts, are bulged at the groining levels, and lean out towards the court. The groining is split along the soffits in almost every gallery except one, where alone iron tye-bars across the arch-chord (originally placed in all), remained, after the French division that was quartered here; a proof of the value of such bars in seismic construction, as well as in the eyes of the brigands that pillaged them.

Opposite the front entrance gate of the monastery, at D, (Diagram No. 240,) but at a considerable distance to the south beyond what the diagram admitted, stands a monument to St. Bruno (Photog. No. 225). The general plane of the structure runs east and west almost exactly. Several of the little obelisks and finials upon it, have been twisted from left to right (looking south at it), or in the same direc-

THE MONUMENT IN HONOR OF ST BERNARD.

GREAT

LIB

LAR

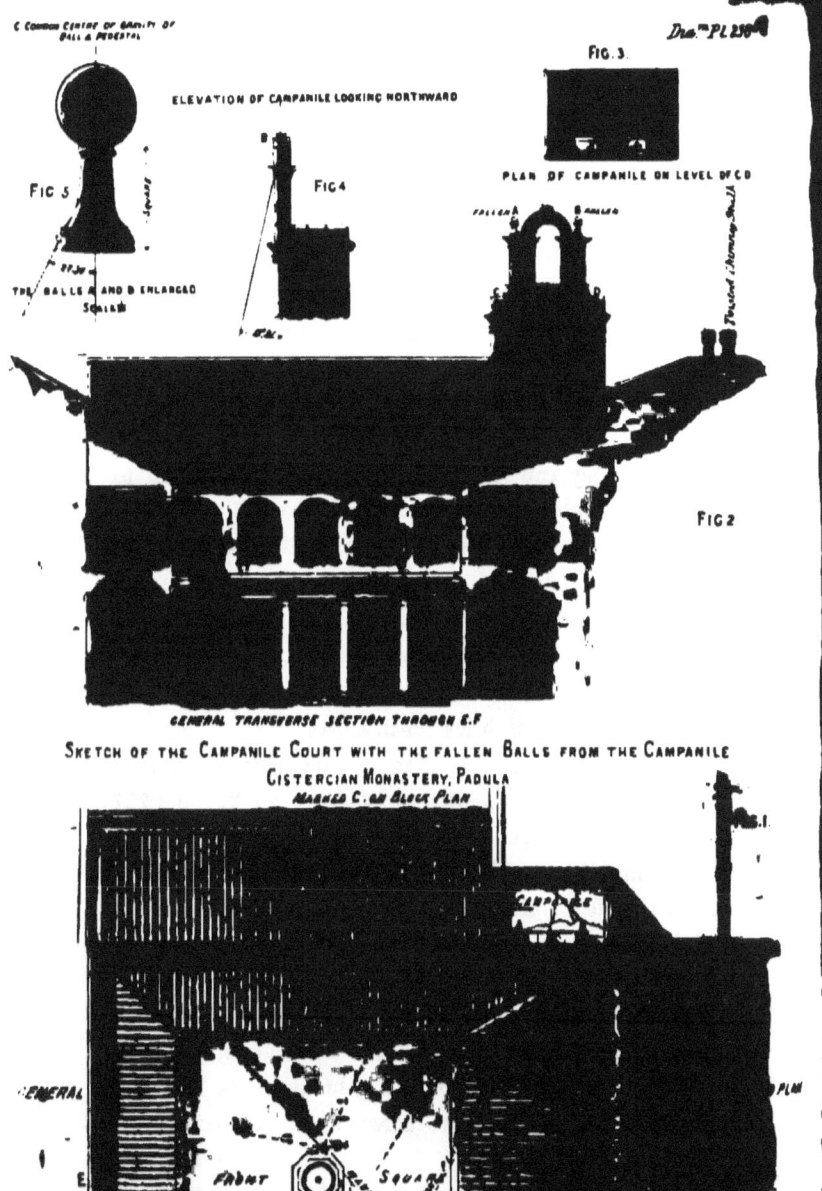

tion as the hands of a watch move. Some of its smaller are thrown down.

In the entrance square, within the walls, as seen in Photog. No. 226 (Coll. Roy. Soc.), looking south from the steps S (Diagram No. 240, Fig. 1), the whole of the square pyramidal chimney caps upon the east side building, from B to B, are thrown down, and on to the roof tiling; and several of the balustrades and finials over the mural fountain, in the centre of the length of this side, are twisted also.

The great axial line of the whole mass of buildings is north 5° W.

The chimney caps B B, 13 in number, are thrown to various horizontal distances, varying between 3 and 5 feet, on to the tiling, as in Fig. 2 (Diagram No. 240), but all with a very nearly uniform direction of 136° E. of north. None of the chimney shafts, which stand about 5 feet above the roof, have been overthrown, though some have been shattered at top by the chucking off of the caps, which, like the shafts, are of brick set in mortar.

Entering the front court or square, the campanile is over the south-east corner, as seen in Photogs. No. 227 and 227 *bis* (Coll. Roy. Soc.), and from the summit, two large balls, of stone, A and B, have fallen, one of which, with its pedestal A, remains buried amidst the broken parts of the roof upon which it descended, while the other, B, falling without its pedestal (which still is *in situ*), upon a more solid part of the roofing, rolled down over the eave, and described a trajectory from B to B^2. (Figs. 1 and 2, Diagram No. 238), down into the marble-paved court, breaking the nosing of the step at B^3, rolling thence across to B^4, and striking the base of the fountain, "made a cannon" over to B', where it remains.

In the north-west corner of this square, beneath the arcades of the west side, a statue of the Madonna, standing in a niche, has been twisted on its base, and shifted in a final direction, 115° E. of north, as seen in Photog. No. 228 (Coll. Roy. Soc.), and Fig. 3, Diagram No. 240.

Very near the Campanile, but far below its summit level, upon the east gable of the front range of buildings, at F (Fig. 1, Diagram No. 238, and Fig. 4, Diagram No. 240), are two remarkable chimney stalks, one of which has been twisted upon its base, at a horizontal fracture close to the level of the gable, the other standing uninjured.

To the eastward of the front square, the roofs of the church, and of the grand refectory, groined and arched buildings of great magnitude, have been heavily fissured. In the refectory both end gables, (Fig. 239,) run up originally against the ends of the brick vault, have parted off from it, and pre-

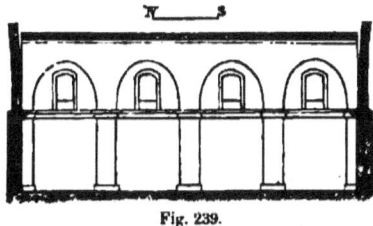

Fig. 239.

sent east and west fissures, open 3 inches at top on the north, and 1¾ inch at the south end, while longitudinal fissures of half an inch to one inch wide run north and south along the soffit. The roof of the church is still more dislocated; both roofs are brick vaults. At the upper end of the refectory the great fresco by Elia, of the Marriage of Cana, is fissured and nearly destroyed. Photog. No. 229 (Coll. Roy. Soc.) shows this, and the great vault fissure at the end.

In the Priors' Square, one of the most richly decorated portions of the monastery, the whole of the buildings are in a falling state, the vaulting of the surrounding arcade being split on each side of the square, the north and south sides, most formidably, and next to these the east side, the front pilasters and the story which they carry above them, all leaning out, and heavily fissured, as in Photogs. No. 231 and No. 232 (Coll. Roy. Soc.)

Entering the great square to the north-east of the last, which in magnitude rather seems the "place d'armes" of some immense "caserne," than the court of a monastery, the groined arcades are similarly fractured; the pilasters and story above, bulging out into the court, are seen from within, and from without, in the line of one of the galleries in Photogs. No. 233 (Coll. Roy. Soc.), and No. 234.

To the westward of the great square, in the private garden of the "Priure," amid much other destruction, a limestone vase has been thrown, from the summit of the south pier of the gate at the west side of the garden, as seen in Photog. No. 230 (Coll. Roy. Soc), and in Figs. 1, 5, and 6, (Diagram No. 240), the direction of throw 122° E. of north.

The blocks of stone of the pier itself, have been thrown or shoved upon each other, eastward about half an inch, and the whole of both piers more or less dislocated.

I have thus briefly recapitulated the objects to be specially referred to, for the information they convey. The vast mass of buildings, however, presented almost an unvarying spectacle of destruction—few of the walls or roofs actually prostrate, but everywhere fissured, dislocated, and tottering;

all their beauty, and magnificence of architectural form and coloured decoration still addressing the eye, but along with gaping rents, that sadly told that their glory was departed; if repair were possible the vastness of the cost precluded it; and thus in a few years hence, the work of one terrible hour, will have made the owl and the bat, the tenants of this Cistercian palace.

Almost the only part of the edifice that has escaped serious injury, is the grand elliptic staircase leading to the park, at the extreme northern end. This noble work, said to have been constructed from a design by Buonarotti, is built of fine sawed ashlar, in the hard white Apennine limestone, everywhere polished withinside. Its preservation seems to have arisen from its form, the support to the north given by the broad flights of steps within and without, and the careful nature of its workmanship.

CHAPTER XII.

FIRST DEDUCTIONS FROM FACTS OF THE CERTOSA—
DOUBLE SHOCKS.

I now pass to the deductions to be obtained from the observed facts here.

There is evidence everywhere, of a double if not a triple shock, confirmatory of the statements made at the town of Padula, of oscillation in various directions. The main shock was in the primary wave-path, right along the Vallone 15° W. of north towards the south, and arrived, through the deep clays and loose material of the plain. This was *preceded* at a very brief interval by a secondary shock, transverse in path to this by a certain angle, and derived from the lateral vibration of the mass of limestone mountain on the range to the north-east. Lastly, the primary shock appears to have been reflected, from the abrupt neighbouring mountain further south, and to have returned again, as an *earthquake echo*, through the clays, with very diminished force, arriving last upon the scene.

Referring to the Photog. No. 225, of the monument of St. Bruno, it will be seen that many of the obelisks and finials are twisted, and some are overthrown. We have universal evidence of the shock, in the path 15° W. of north to south. Here the finials which are *overthrown* are thrown directly westward. All those that are twisted are turned from left to right.

Now there are two distinct trains of earthquake causation, by either of which bodies may be twisted on their bases. 1st. By the action of a *single* shock, when the centre of adherence of the base of the object, lies to one side or other of the vertical plane passing through the centre of gravity, and the line of the wave-path. 2nd. By the conjoint action of *two closely successive* shocks. By the first shock, the body is tilted up from its base, but not overthrown, so that for a time greater or less, it rests wholly upon one edge of its base; while thus poised, if another shock bear upon it, in any direction transverse to the first, it acts as usual at the centre of gravity of the body, to displace it by inertia, in the contrary direction to the wave transit; but the body is held more or less, by friction *at the edge momentarily in contact* with its support, and there only; but this edge must always lie to one side of the vertical plane passing through the centre of gravity, in the direction of the wave-path: hence the tilted body, *while relapsing upon its base also rotates*, round some point situated in the edge of its base upon which it had been tilted, and thus it comes to rest in a new position, having twisted more or less round a vertical axis.

If the observer look due south at a square pyramid, for example, whose sides stood cardinal, and it be tilted by the *first semiphase* of a shock from east to west, the pyramid will tilt or rise upon the eastern edge of its base; and if, before it has had time to fall back, it be acted on by another shock from north to south, the pyramid will rotate, upon the bisection or on some other point, of the edge on which it momentarily rested, and will hence come to repose, after having twisted from left to right, or *with* the hands of a watch.

If the tilting up, had been produced by the *second semiphase*, of the same shock from east to west, then the pyramid would have risen upon the western edge of its base, and the *same* direction (north to south) of second shock, would have produced rotation upon that edge, but in a *contrary* direction to the preceding, or from right to left, or *against* the hands of a watch.

Again, if, on the first supposition, the *first semiphase* of the east to west shock, had tilted the pyramid upon its *eastern* edge of base, but the second shock had been from south to north, in place of the reverse as before, then the rotation would have been from right to left; and if tilted by the *second semiphase* on the *western* edge, the second shock, south to north, would produce rotation left to right.

It would therefore appear at first impossible, to determine the *direction* of motion in transit, of either shock, from such an observation: we can, however, generally discover upon which edge of the base any heavy body of stone or masonry has tilted, by the abrasion or splintering of the arris, and the rotation must have taken place round some point in that edge. If, therefore, we know the direction of either one of the two shocks, we can always discover that of the other, by the rotation observed; and if the time of oscillation of the body be ascertainable, we are enabled to calculate a major limit, for the interval of time that must have elapsed, between the arrival at the twisted body, of the first and of the second shock, when both the wave-paths are known.

With a single instance of such twisting, it may be impossible to decide, whether the twist has been due to one shock, (1st case) or to two shocks in succession, (2nd case); but when several bodies alike or dissimilar, at the same locality, are *all found twisted in one direction*, it is certain to have been *the work of two distinct shocks*, for it is beyond the reach of probability, that several bodies, should *all* happen to have their respective centres of adherence, at the *same side* of their respective centres of gravity, and unless they have, some will rotate in one, some in the other direction by any single shock; rotation thus produced, being always by the centre of gravity, moving contrary to the first or second semi-phase of the wave, and carried round the centre of adherence, by the line joining them as a radius vector; the inertia of motion at the centre of gravity, and the resistance of the point of rotation in the edge of the base, or of the centre of adherence, forming in every case, the extremities of the dynamic couple.

All the effects of the double shock will be understood by examination of the Figures Nos. 235 and 236, in which

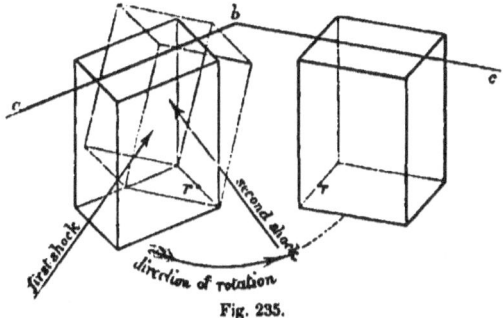

Fig. 235.

Fig. 235 shows the action of any double shock; Fig. 236 the variations of result produced, first, by rotation in the first semiphase A, and second semiphase B, by the same double shock; secondly, the like by rotation, in the first semiphase (C), the first shock being as before, but the second, contrary in direction to that of the previous cases (A and B), and the like for the second semiphase (D), the two shocks being the same (C and D).

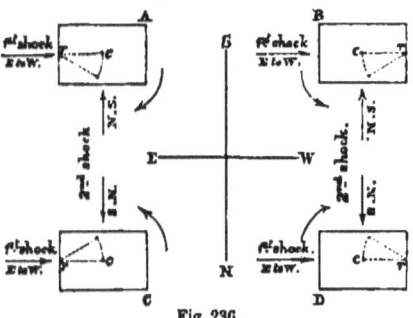

Fig. 236.

Applying this to the facts at the monument of St. Bruno. *All* the finials, &c., are twisted from left to right; we know that the main shock was from 15° W. of north to the south, it therefore follows, that the shock which first moved them, arrived in a path somewhere between that, and from east to west: by this they were *tilted*; by the immediately following shock, 15° W. of north to south they were *twisted*. Neither shock was sufficient, in velocity or range, completely to overthrow any of them, except those which were top-heavy, by having had balls at their summits, which have, except in one instance, been all dislodged.

A great many pyramids and finials on the top of the fountain in the entrance square B, Fig. 1, Diagram No. 240, and Photog. (Coll. Roy. Soc.), presented precisely similar phenomena, as did those on the parapet of the great façade Photog. (Coll. Roy. Soc.), and in divers other places.

The complex forms of these objects, which rendered the ascertainment of the positions of their centres of oscillation on the edges of their bases, difficult and uncertain, unless by experiment, prevents any calculation of a precise character, from their movements, as to the velocity of either shock, nor do we require it.

They give us other valuable information, however. In the case of the parallelopipedal chimney, (F. Fig. 1, Diagram Nos. 238-240, and Fig. 4, same diagram), twisted upon its base, it had rotated upon a point in the western edge of its base at b, Fig. 237. We know already that the direction (generally) of the first shock, was from some points east or N.E. towards the west or S.W., the second being from 15° W. of north to south. The chimney stalk had therefore made, *one semi*-oscillation, and *one complete* oscillation; that is, it was being acted on by the *second* semiphase of the wave of the *first* shock, at the moment when the second shock arrived at it, as in Fig. 237.

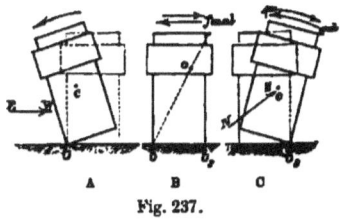

Fig. 237.

The centre of oscillation of the chimney above b thus tilted was, as nearly as could be ascertained, 4·33 feet distant from the edges of the base upon which it tilted b and $b_{,}$. The first shock, east to west, fractured the chimney from its base, and produced in the detached chimney, one semi-oscillation eastward (A, Fig. 237). The chimney then relapsed upon its base (B), Fig. 237, and rising again upon the edge b_n leaned over westward (C), Fig. 237, having thus made one complete oscillation in that direction, with the moment of repose (B), when it had fallen back plumb upon its base. Between that moment of repose, and the completion of the oscillation, or *almost instantly after it had commenced to fall back* (C) from west to east, to reassume its original position of repose, the second shock from the north to south reached it, and twisted it round horizontally, in the manner that has been already explained.

CHAPTER XIII.

DEDUCTIONS FROM FACTS PRESENTED AT THE CERTOSA, CONTINUED—INTERVAL OF TIME BETWEEN THE FIRST AND SECOND SHOCKS CALCULATED.

We can calculate, therefore, from these data, to a good approximation, the interval of time that elapsed, between the arrival of the first and second shock, and thence the difference in transit velocity, of the two waves of shock; the first through the limestone, the second through the deep clays and gravels, &c.

The chimney may be regarded as a parallelopiped vibrating as a compound pendulum, upon b, and b_2 (Fig. 237) as points of suspension, whose centre of oscillation is in the plane of shock passing through the centre of gravity, and distant from b or b_2 by two-thirds of the diagonal of the parallelopiped, in the same plane.

This distance $= 4\cdot 33$ feet is the length of the corresponding simple pendulum, with some small allowances for the hollowness and irregularity of form. The greatest possible arc of vibration, is limited by that which would bring the centre of gravity, c, vertically over b and b_2, Fig. 237, beyond which the mass must have fallen. This I found to be 21°, at either side of the vertical through the centre of gravity, when the chimney was in its original undisturbed state.

The arc actually described, must have been less than this,

and from other observations yet to be referred to I conclude, that this arc was not more than 15° at either side the vertical, or with a horizontal chord at the centre of gravity of about 12 inches. We shall not make a sensible error, however, by assuming the arc described, as the largest possible.

Taking the time of oscillation from the equation

$$T = \pi \sqrt{\frac{l}{g}} \times \left(1 + \frac{a}{8\,l} + \frac{9\,a^2}{256 \times l^2} + \ldots\right)$$

where $a =$ ver sine of half the arc of vibration; we have

$$T = 3\cdot 14 \sqrt{\frac{4\cdot 33}{32\cdot 2}} \times \left(1 + \frac{\cdot 066}{8 \times 4\cdot 33} + \frac{9 \times \cdot 066^2}{256 \times 4\cdot 33^2} + \ldots\right)$$

or $= 3\cdot 14 \times \cdot 3668 \times 1\cdot 0019$ approximately,

and $T = 1\cdot 536$ seconds;

correction for latitude being needless.

From the moment of arrival of the first shock, up to the arrival of the second transverse to it, the chimney had made, one half and one complete, vibration, and possibly had just commenced another. At the moment that the chimney relapsed upon its base it lost *vis vivâ*, and therefore *time*, before it rose again, to complete the arc westward This minute loss of time we can only estimate, because although we know the velocity at the moment the mass of the chimney struck its base, on resuming the perpendicular (B, Fig. 237), we do not know its hardness, elasticity, &c., upon which that loss also depends.

Neither can we calculate precisely, how much of the commencing arc of the third oscillation had been performed (if any), before the second shock reached the chimney, because we do not know the precise point round which it rotated, &c.

But where the arc of horizontal rotation is great, as it is in this instance, where the chimney was twisted round 30° to the westward, the latter must have been extremely small or *nil*.

Calling these two small corrections d and f, the whole interval in time I, between the shocks is

$$I = T + \frac{T}{2} + d + f.$$

We therefore obtain

$$I = 1·536 + \frac{1·536}{2} + d + f.$$

or without the two latter,

$$I = 1·7304 \text{ seconds,}$$

and estimating d, assuming f as small $= 0$, we may consider

$$I = 1·75 \text{ seconds.}$$

This is the difference in time, between the arrival of the two shocks; both of which started at the same moment and from the same origin. The evidence for the position of the latter has yet to be adduced; anticipating the *fact*, however, the Certosa, is distant from it, $16\frac{1}{2}$ geographical miles, or 33,415 English yards in the right line, which was the path of the second shock, arriving directly 15° W. of north to south, through the deep clays and gravels, &c., of the Vallone; while that of the first shock, was through the east lateral range of solid limestone mountains, but whose length (as respects the wave-path) we may consider as one-fifth of a mile greater.

From facts (also yet to be adduced), I found that the general velocity of *translation* of the wave of shock, through the limestone country, was at the rate of 240 yards per second. This, therefore, may be taken as the velocity of translation of the first shock here (through the limestone); the total time of its transit from the origin (surface velocity) is therefore

$$\frac{33415}{240} = 139 \cdot 230 \text{ seconds};$$

and through the clays and gravels the whole time is

$$139 \cdot 230 + 1 \cdot 750 = 140 \cdot 980 \text{ seconds}.$$

The velocity per second of surface translation in the clays and gravels was therefore

$$\frac{33415 + 402}{141''} = 239 \cdot 84 \text{ yards per second}.$$

The shock through the limestone reached the Certosa from the Colline of Padula, the nearest point of rock in the path, through an intervening stratum of clays and gravels, by which there must have been some loss of *vis vivâ*. If we throw off the decimals for this correction, which we can only estimate, we have finally, for the surface velocity of translation, 240 yards per second in the limestone, and 239 yards per second in the clays and gravels, measures approaching so nearly to equality, as to warrant one or both of the following conclusions.

Either, the main primary or direct wave arrived, like the other through the limestone formations, deep beneath the clays and gravels of the piano, and shook the latter resting upon them, at their own rate of translation nearly; or (here at least), the bedding-joints and other breaches of homogeneity in the limestone rock produce a retardation of the wave transit therein, such as reduces the velocity to nearly that in the dense clays and gravels. In this extremely dislocated and overturned country I deem the latter as most probably the fact.

The measures of velocity for this locality are relatively, however, as nearly reliable as the data will admit. As absolute measures of transit velocity, they must be taken as mere approximations, as all our data for this are based upon the observations made as to time at the moment of shock, and unfortunately are not only few in number, but by no means to be relied upon as to exactness, in this locality.

CHAPTER XIV.

DIFFERENCE IN AZIMUTHS OF PRIMARY AND OF SECONDARY SHOCKS AT CERTOSA— ANGLE OF INTERSECTION.

I NOW proceed to consider the angle of intersection in an horizontal plane, made by these two shocks at the Certosa. Referring again to the two great stone balls (one along with its base) projected from the top of the Campanile upon the roofs of the front square, (Fig. 1 Diagram No. 238 and Diagram No. 240, and Photog. No. 227 (Coll. Roy. Soc.) in which the ball B is seen lying, where it fell), we are enabled from these to infer the wave-path of the first shock with some certainty.

Both balls were projected from the Campanile in a direction 64° W. of north, and precisely alike.

The first shock, arriving from somewhere between north and east, caused the Campanile to oscillate from west to east and east to west; *i. e.* in the line of its narrowest dimension of base.

The balls were thrown off in the second semiphase of the wave; therefore, and as the time of oscillation of the tower (from its altitude) was large, and greater than that of the whole phase of the wave they were projected with a velocity *less* than that due to the shock; the top of the tower moving, by its elasticity and inertia, towards the east, at the same time that the forward movement of the whole mass towards the west, by the wave of shock, projected the balls with a velocity equal the difference.

We shall take the velocity from the ball B, which fell unencumbered by its base, and whose mark where the sphere struck and fractured the tiling of the roof, at B, (Fig. 1, Diagram No. 238, and Fig. 2, Diagram No. 240,) left no doubt as to its precise range.

This ball was projected a horizontal range of 12 feet, along the plane of projection, and descended vertically 42 feet from the summit of the Campanile to the roof tiling.

We may assume the angle of emergence for this shock $e = 20°$, the same as that of the primary shock (15° W. of north to south) here; because although coming from the limestone in which (at Padula) we found the emergence greater $= 25° 30'$, yet we are here some hundreds of feet lower, and the shock, in passing from the limestone into the clays and gravel, intervening before reaching the Certosa, must have suffered some refraction into a rarer medium, both tending to reduce the value of e.

Both these balls were attached to their pedestals or bases by a small wrought-iron dowal inserted into both, sufficient in strength to communicate its velocity from the base to the ball, but permitting easy separation of the two (*i.e.* ball and base) when not rusted into the sockets. It was so rusted, in the ball A which fell with its base, but in B, the dowal, which was about half an inch diameter, was found broken off, either previous to, or during the fall of the ball.

It must be borne in mind, in considering what follows,

that these balls, and more particularly that B, were *projected* by the transverse shock, the impressed movement being therefore in the *same* direction as that of the wave, but had their plane of projection altered by the immediately following main shock, which, acting on them by inertia, impressed a movement in the *contrary* direction to that of the wave. Unless this were understood the path obtained by the following method would appear erroneous.

We obtain the velocity of projection from the equation

$$V^2 = \frac{a^2 g}{2 \cos^2 e \, (b + a \tan e)},$$

in which $e = 20°, b = 42$ feet, $a = 12$ feet,

and $V = 11·54$ feet per second.

This was the velocity in the plane of projection 64° W. of north, but this direction was not that of either shock but a resultant of both, the ball having necessarily received a certain amount of impulse from the first shock, and before it had completely parted hold from its support, been exposed to the impulse of the second. Now the direction of the second (*i. e.* the main) shock was 15° W. of north to south, and we have already ascertained at several points that its velocity was 12·97 feet per second. We have, therefore, two velocities, and the direction of the resultant, and of one component given, to find the direction of the other component.

Resolving, we find that the transverse shock made an angle with the primary or main one of 56° 40′, and that the wave-path or direction of the former was 41° 30′ E. of north to south, which is precisely the direction of a line drawn from the extremity of the mountain range northward and eastward of Padula and to the Certosa, as may be seen by examining Zannoni's great map (Sheet 19). The range, after running nearly north and south, terminates at Padula, after having made an abrupt bend to the westward in the above azimuth (by compass) from the Certosa. Hence it is to be inferred that the transverse shock (the first in point of time) was delivered from the *free* extremity of the range of limestone mountain, along the axis of which it had followed, while the great primary or direct shock (the second in point of time) came normally through the clays, gravels, and other formations, beneath and in, the Vallone di Diano, both intersecting at the Certosa.

CHAPTER XV.

FURTHER DISCUSSION OF OBSERVATIONS MADE AT THE CERTOSA AND AT PADULA.

At Padula, while the main wave-path was unmistakably 17° W. of north to south, I found fissures giving extremes of directions from 155° E. of north to 166° E. of north, and evidences of a very subordinate vibration nearly from west to east. The latter, it is highly probable, was due to the partial *dispersion* of the main wave of shock, as it reached the southern head of the great valley, and passed from its deep formations out into the limestone mountains that shut it in to the south.

I had also evidences of such subordinate movements, and more distinctly marked, at the Certosa.

The great fissures here, by the very construction of the buildings, ran principally east and west, and north and south; the former being by much the wider (transverse to the main shock), but the abutting of the several masses of building upon each other, very generally preventing that freedom of motion, that is essential to enable deductions as to wave-path to be made thus with precision. From some of the main buildings, that rose free and unencumbered above the level of the surrounding ones, however, I obtained measures of direction, the extremes being 116° E. of north to 165° E.

of north; and although very wide in limit, and often perplexed, from the complicated movements to which this place had been subjected, all were confirmatory of the general wave-path obtained at the Francescani already given.

The most instructive fissures I found, were those by which the two great gable walls of the refectory had parted off from the ends of the semicircular vault which formed the roof. An interior Photog. of this noble hall is given (No. 229, Coll. Roy. Soc.). The soffit of the vault was fissured in a north and south direction from end to end nearly, and half to one inch wide.

The gables were great semicircular walls of rubble stonework, run up from the chord-line or level of the springing, after the vault had been turned, and merely closed in against the ends of the arch, as in Fig. 239, not being bonded into it. Each of these gable walls had gone out at top from the end of the vault, producing a gradually widening and tremendous fissure, which, at the north end, was open 5 inches at top, and at the opposite, or south end, 3 to 3½ inches.

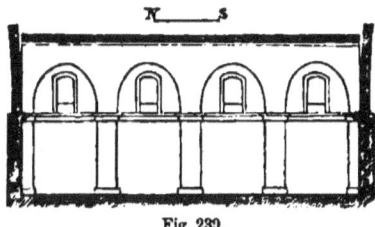

Fig. 239.

The gables, thick, heavy, inelastic, and yielding, from the bad rubble of which they are built, had bent over from above the chord-lines, showing innumerable minor thread-like fissures of dislocation in the work, more or less horizontal; neither had approached within some inches of the limit of stability. The north gable wall had given out in the first semiphase of the shock, the south in the second semiphase; and both appeared, from the fragments fallen into the fissures, none of which were crushed, or gripped and pressed into compacted powder, (as is not unusually the case,) to have gone out very little, if at all further, by the impressed movement, than where they stood as I examined them.

The width of this north fissure, therefore, affords an approximate measure, of the actual amplitude of the main wave of shock here; for the opening, or the actual movement of the gable, at the level of its centre of oscillation, upon the chord as axis, must have been about equal to the amplitude of the wave, taken in a horizontal line. It would be useless, upon inexact data, to pursue this minutely; we may conclude, however, that the horizontal amplitude of the main wave of shock here, did not much exceed 4 inches.

There were also immense fissures, in the 9-inch brick groining of the roof of the church, both longitudinally and transversely through the axis. The beautiful domed cupola, also presented complicated fissures, but was in so tottering a condition as to prevent close examination; all these indicated a principal wave-path some degrees W. of north to south.

The urn or vase, thrown from the summit of one of the gate-piers, in the garden of the Priure, Photog. No. 230 (Coll. Roy. Soc.), and Figs. 1, 5, and 6, Diagram No. 240, presented a good example, of the high velocity with which bodies thus placed on the summits of slender vertical erections, may be projected by the elastic vibration or rocking communicated by the shock to the erection itself. This vase was thrown nearly in the same path as the balls from the Campanile, but in the opposite direction. Its path was 58° W. of north to south.

It was, no doubt, thrown off by the recoil of the pier upon which it stood, when making its *second* oscillation, produced by the first shock (41° 30′ E. of north to south), when the main shock reached it, which increasing the movement already impressed, and at the same time changing the plane of projection, threw it to an horizontal range of 18 feet, from an elevation on the pier of the same (18 feet).

Assuming $e = 20°$ as before, we have from the equation

$$V^2 = \frac{a^2 g}{2 \cos^2 e \,(b + a \tan e)},$$

the velocity of projection of this vase,

$$V = 21 \cdot 23 \text{ feet per second},$$

which is about 8¼ feet per second in excess of that of the wave of shock, the difference being due to the angular velocity of elasticity acquired by the pier itself, *added* to that of the wave. In fact, in such examples, the body thrown is projected like a stone from a sling.

There was every reason for my believing that the vase remained where I found it, half embedded in a flower-border, in the locked-up garden of the prior, and untouched since the earthquake; and the splintered and disjointed state of the limestone blocks of the pier, itself indicated the extent to which it had vibrated. The corresponding vase was not thrown though loosened, nor was the pier as much shattered, though I could not discover any very certain cause for the difference in effect upon both. Each vase had been steadied when in place by a slender iron dowal.

I have stated that there was evidence of a *third* shock, different in direction from either of those already considered. This was visible in many small objects, which gave indications of disturbance, by the main shock or by the transverse one, and also of immediately subsequent disturbance, by another, or by several other minute vibrations or little shocks, in directions from south to north, varying 10° to 15° to the east or west of that. This last shock, or jarring succession of shocks, appears to have been a true *earthquake echo*, or reflection of the main shocks back, from the limestone mountains to the S., S. W., and S. E. of the Certosa.

In the line of buildings to the east side of the great front entrance square, between B and B (Figs. 1 and 2, Diagram No. 240) a great number of pyramidal brick chimney caps were thrown off from the tops of the stalks, in a general direction to the S. E. The *mean* direction of their throw I found to be 136° E. of north, which is one not so widely different from that of the resultant path of projection due to the two main shocks, but that all of these might have been projected off at the same moment and by these shocks; the differences in direction being due to the irregular figure of the pyramids, and to their ordinal position with reference to the resultant path, as well as to their having in some cases probably slided after their fall upon the sloping tiling of the roofs. The position of several of these caps, however, and the wide diversity from the resultant path of others, caused me to conclude that several of them had been loosened by the main shocks, and afterwards overthrown by this third movement in reverse.

To the same repetition of movements I attributed the singular displacement of the limestone statue of the Madonna (Photog. No. 228 (Coll. Roy. Soc.) and Fig. 3, Diagram No. 240), which I found had moved upon its pedestal, without injury or overthrow, about $1\frac{3}{4}$ inch in a direction 115° E. of north towards the S. E. Its base was an irregular octagon. The figure had been twisted a little in a direction from north towards east, or with the hands of a watch; and its displacement appeared to have arisen, from its having rocked like a conical pendulum, round the successive sides and angles of its octagon pedestal, which the base of the figure overhung: by the conjoint influence, of the intersecting shocks, and the centre of gravity of the figure being not over the centre of the base, but nearer to the S. E. side (at which the infant rests in the Madonna's arms), the circle of gyration tended to this side, and as the figure passed round each angle at that side of the pedestal, it gained a little ground towards the direction in which I found it had shifted. The friction between the pedestal and base, there being no cement and both smooth, being small, it would be possible for a figure of this sort, however, to shift its position by merely rocking to and fro in one plane at first, the lower part shifting forward at each return oscillation, by the existence of a centre of "spontaneous rotation" between the centre of gravity and base; and such may have happened here, complicated by the two nearly concurrent intersecting shocks. The centre of gravity appeared to the eye, to be at about 2 feet 2 inches above the base, and the weight of the statue was about 3 cantari, or $6\frac{1}{2}$ to 7 cwt., but I had no means of further examination. The Priure, and still more the Vicario, Il Padre Bruno Santullo, were enabled to give me an intelligent and consistent account, of their experience and observations during the earthquake. They, and all with whom I conversed in the monastery, described the noise as being heard at the same moment, as the first movement of shock; some thought a *very* little before, some an instant after, and that it continued as an awful rumbling roar during the whole time of motion, and even after it. They could give but a very confused account of the second shock, which arrived about an hour after the previous ones; saying that they were all in too much alarm, and dread of the buildings around them, that were momentarily giving signs of falling, to be prepared to remark much about the last shock, except that it further ruined and shook down many things that the preceding ones had left. They had all taken refuge in the centre of the great open court, and remained exposed to the cold for several hours, before they durst return to their shattered cells.

They were unable to give any precise information as to the moment of occurrence of the first shock; and as to direction, they could only say they were shaken in *every* direction, and that the shocks were at first, they thought, from north or N. E., and then in every way or vorticoso. The only watch in the monastery, was a curious old English one, the maker's name and date in which proved it to have been made about 140 years before; and with a singular notation for the hours upon the dial, which the owner had never been able to make out until I deciphered it for him. It may therefore be imagined that it was not kept very exact as to time, and probably had not for a hundred years shown true time, until the day when I set it at noon, for the venerable and simple-hearted owner.

CHAPTER XVI.

PADULA TO MOLITERNO, BY THE PASS OF ARENA
BIANCA AND LAGOMAOURI.

I LEFT my kind hosts of the Certosa at early dawn, on the 16th February, to pass still further south in the valley, and ascending over the pass of Ovedone, or Arena Bianca, to reach Moliterno, in another valley to the south-east.

Approaching Sassano, a town on the west side of the valley, four miles from Certosa, I was enabled to see with the telescope that it had suffered but little, as was also the case with St. Giacomo, to the west of it. One building, partly ruined, and above the level of others, I could see distinctly, and by the fortuitous circumstance of the light from the sun reflected from its window-panes, whose azimuth I observed, was enabled to calculate approximately the axial line of building, and to infer the direction of the fractures and of the shock there. It turned out to be 137° E. of north. I cannot lay much stress on an observation so made. The muleteers, who knew the place well, said the people of it believed the great shock to have been from north to south, and that the same was the case with Buonabitacola, a small village at the same side, but further south. I passed near the latter, and the muleteers pointed out, where some large fissures had been produced in the earth near the road. I could not spare time to visit either, and commenced the ascent towards Ovedone, the mules creeping up, by stony traverses, along the N. E. side of the Fiume Imperatore, a torrent falling into the Calore, on its right bank. The section of the mountain range crossed, which separates the valley of the Calore, from that of the Moglia and Agri, and the general features of its geology, so far as I could observe them along my mule tracks, and thence on to and beyond Montemurro, are given in the section, Diagram, No. 241, Fig. 1.

After three hours' ascent, the form and features of the surrounding amphitheatre of mountains, by which the Vallone of Diano is shut in southward, became well displayed. To the south-west, stretching away above two miles, the Bosco della Cerzeta, is beneath me. The little town of Casalnovo, which is said to have suffered but little, is just visible beyond it, and high above, at some ten miles away, the gigantic Monte Cocuzzo, and further west Monte Rotundo, with Monte Cervaro due south, both of massive grandeur in outline, rise above innumerable lower peaks and hills, and shut in the view. I am 3,000 feet above the sea, but still these lofty summits subtend a considerable visual angle above my horizon. We are here upon the southern edge of the region, of actually ruined and overthrown towns, the meizoscismal, as reference to the Map, B, will show, all those further to the south being merely fissured, more or less severely. And from this elevation, it is easy to see and understand the physical features of the country, that have produced this sudden reduction of effect, by the prodigious loss of *vis vivâ*, that the wave of shock coming south must have sustained, by the abrupt lowering of the east range of mountains of the Vallone di Diano, almost amounting to a local extinction of the ridge at Padula; as well as by its entire change of direction, and breaking up into numerous detached masses, after passing the low lying intervening gap.

The residual wave entering the new mountain system, has still had power to do great mischief. Casalnovo, Sanza, Casella, Podaria, Le Celle, Montano, Laurito, Policastro, Lagonegro, (where I learned at the Certosa facts proving that the wave-path was there north to south), Rivello, Bosco, Lauria, (where the wave-path was also north to south), Trecciena, Maratea, Tortora, Ajeta, and down to Casalito twenty-seven Italian miles south of where we stand, and innumerable other smaller places between, have been more or less shattered, though no lives have been lost south of Montesano.

The latter town is high and close above me, a little to the south, perched on the crest of a conical hill, well buttressed, and connected with the mountain ridge and shoulder on which it stands beetling to the north. As I pass, close and beneath it, I find from some people of the hamlet of Arena Bianca, (a little to the north), that a few houses and other buildings, and three churches, have been partly or wholly thrown down; and in the clear morning light, with the telescope of the theodolite, I can observe the fissures of many of the buildings, and, by the aid of the compass, approximate to the *path* of the wave, (though neither its *direction* nor emergence). I judged it to be from 150° to 165° E. of north. Near as Montesano seemed, I found it would lose 2½ hours to climb up to it.

The brilliant sunlit dawn, gradually got overcast; a

strong wind from the N.W. sprang up, and the remainder of the day's journey, and that of three succeeding ones, was made under torrents of unceasing rain, which swelled the mountain streams, and the great rivers besides, and rendered their passage occasionally perilous to the laden mules.

At Arena Bianca, from which the pass over the shoulder south of Monte della Vajana takes its name, though called indifferently that of " Ovedone," two houses are down, and numberless breaches are visible, by lengths of the drystone fence walls which abound here, (like those in the limestone country of the west of Ireland,) having been prostrated; almost all the walls down ran east and west; a few, however, had run north and south, and afforded evidence, that while the main wave-path was still nearly north to south, there was here, also a minor transverse shake.

CHAPTER XVII.

JOURNEY OVER THE PASS AND BY LAGO MAORNO.

As we had ascended, the soft, ill-bedded limestone of the lower roots of the hills, had gradually given way to indurated liassic-*looking* hard, clastic, cherty beds of yellow limestone, in distinct and highly-inclined layers, probably metamorphic (see Section)—these, after many alternations came back to the same soft, cretaceous-looking stuff, and are finally lost, (as we top the steep and get upon the more level table of the shoulder and elevated valley), under a deep loose deposit of almost perfectly pure and fine white sand, (from which the pass and town derive their name), mixed largely, but unequally, with impalpable chalky particles. This extends to beyond Tardiano, under the southern summit of Monte Vajana; but at various points the rock beneath might be seen, and proved that we had lost the limestone proper, and got upon an endless succession of thin beds, all more or less metamorphic, consisting of ash-yellow limestone, often highly quartzose, hard and clastic, alternating with thick and thin beds, of grey and purple and green clays, and of variegated marls; and in one place some beds of 4 to 6 feet of impure alabaster, all highly inclined. (Geolog. section, Diagram No. 241.)

At the highest point of the pass, near Arena Bianca,

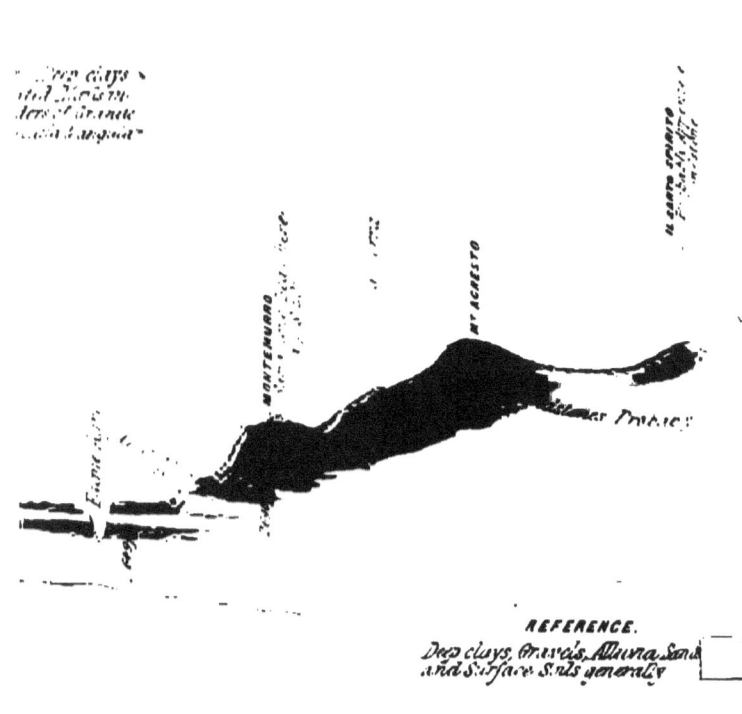

REFERENCE.

Deep clays, Gravels, Alluvial Sands and Surface Soils generally

Calcareous Breccia with residual pebbles of Limestone, Quartz, Slates and Metamorphic Schists and Sandstone

Variegated Marl Beds green, purple grey, including Gypseous and Metamorphic Limestone beds

Sandstone with Argillaceous and Calcareous beds Shales Clay beds and lignites

Metamorphic limestones, indurated flinty and occasionally Dolomitic

Soft chalky and arenaceous Limestone Fossils, Nummulitic and others

Fossil with bedded Limestone Pebbly Serpentine

Section, Fig 1.

the barometer stands at 26·50 inches, thermo. 42° Fahr. at 12·30 P.M., and the elevation above the sea at Naples is 3158·3 feet. At Tardiano, the contrary slope towards the valley basin of the elevated mountain lake, the gloomy Lago Maorno commences. This lake is commonly pronounced Lago Maouri by the peasantry. Many singular beds of metamorphic shales, and indurated clays, with some thin seams as hard as chert or jade, and, lithogically, very like the latter, are passed, and at about 500 feet above the level of the Lago Maorno, where the descent has become again very steep, along by the northern side of a nameless torrent that discharges into it, great beds of intensely hard, flinty, and hæmatitic, dark-blue grey limestone occur, with huge seams of very hard, black and brown hæmatite, 12, 15, 24, and 30 inches thick. All these run here in a general direction, for a considerable way, nearly east and west, but as they extend, are much twisted towards the north. They are very nearly vertical, their dip being only 15° to the south. The torrent separates these beds to the south from the cretaceous limestone, which here makes its appearance once more. Sloping up steeply to the north for more than half a mile, with a bare, weathered, and water-furrowed surface, without a blade of vegetation, extend parallel, and nearly vertical beds of green, grey, and purple clays, alternating with beds of yellow, soft, clayey sandstone, many yards in thickness, standing in ridges high above the worn-down clays. (See enlarged plan and section, Diagram No. 242.) My attention was directed, in particular, to these beds of ferruginous rock here, by finding that the compass would no longer work, the needle turning round 90°, within a few paces' change of position. On examining the hard ferruginous beds, of thick, blue grey, flinty limestone at A (plan and section) I found, with some surprise, that they exhibited at the exposed south face, for about 200 feet in length, numerous large fresh-made fractures, running nearly vertically, right through the whole of the strata, up to the dense but soft clay beds to the north. Further to the east the beds are covered up by clays on both sides, and the ends of the strata alone visible. These were much obscured by the torrents of creamy clays that the rain was washing over them; but yet I was enabled to trace similar fractures across the tops of the beds, and extending vertically downwards through them, for a distance of about 250 yards eastward. To the westward, the torrent runs through these indurated beds, and cuts off those to the west of it from any contact visible with the clays.

At the base of the vertical beds below A, in the bed of the torrent, great masses of freshly-fractured and fallen rock were lying, and in several of the fractures, the widest of which was open, on an average, about an inch, on the south face, of the nearly vertical beds, I found fresh fractured splintering fragments in various spots. These had dropped down a few inches only, and I could replace and fit them into the spots from which they had fallen in the jaws of the fissures. There could be no mistake as to the freshness of the fractures, for all the old and weathered portions of the rock, were a deep iron rust, in colour; but the fresh-broken surfaces, a bright blue grey, of a deep tint. Many large fragments from the tops of the upcrop of the beds, had also been detached at weathered fissures, from the south face, and lay thrown into the bottom of the torrent.

It was unmistakably obvious, that the fractures had been produced by the transit of the earthquake, and that the push, of the vast piled-up mass of comparatively soft heavy clay beds, &c. to the north with which they had been forced against the barrier of these hard limestone beds that ramparted them in, had been of such force as to fracture the latter in many places. The hard unyielding rock had broken; the softer clay beds had merely been slightly compressed, and changed insensibly in form; hence they presented no evidence of the force that they had transmitted to the fractured rock. On ascending the slope of the clay beds, northward about 500 yards, however, I found confirmatory evidence of my conclusion, in great masses of fresh fractured and fallen sandstone from the thick beds at C. (Diagram No. 242.)

The N.E. face of the deep gorge beyond C, not shown in diagram, that brings another torrent down in a N.W. to S.E. direction, also showed great falls, of these soft sandstone cliffs in the bottom. I had great difficulty in descending the wet clay beds, which, devoid of a single pebble, presented no foothold whatever upon their unctuous and slippery slope, and a fall produced much the effect, of being dipped into a succession of paint pots.

This fact was to me one of peculiar interest: it was the first example I had found, of actual fracture of beds of hard rock *in situ*, by the impulse of an earthquake wave—a phenomenon in kind totally distinct, from such *breaking off* of great masses, as the rock falls of Campostrina, or the Arguilles of Padula. Here were beds of the very hardest and toughest rock, such as, with difficulty, I broke specimens from, with the hammer, fractured for many yards in depth (fully thirty yards was visible) and extending over a great length of bedding, and yet free from any other sign of violence, or any other sensible disturbance of position. It realized forcibly to the mind the enormous power of the impulse of shock with even this moderate velocity when acting at once upon great masses, at free or outlying surfaces; and is suggestive of the much more potent effects, that must be occasionally produced in loftier ranges, subject to still more powerful impulses. Hooker's account of the rock dislocation witnessed by him in the Himalayan shocks recurred to mind. Strictly interpreted, however, even this, is but an example of dislocation by secondary effects, of the wave, not by the wave itself.

The Lago Maorno now comes into view, a dreary pool of about a mile long by half as much wide, in the midst of a mountain basin, surrounded with deep tenacious clay soil, apparently of great fertility, but swampy and wet. Across this the mules passed with much labour, sinking nearly to the knees. The shallow valley basin, devoid of tree, house, or human being, is surrounded with low barren hills, all apparently of soft limestone, those to the eastward of the lake, coming down close and abrupt to its margin. Above and beyond the hills to the N.E., Monte Spagnoletto rises high, and powdered with snow; and far away to the S.E., I see the lofty summits of Monte Raparo, and Monte Armizzone, deeply covered with it, as well as the intervening ridges.

All attempt to cross these now, and gain access to Castel Saraceno, Chirico Raparo, Carbone, Calvera, Latronico, Episcopio, and many other towns lying deep in the mountain recesses, far to the S.E. and east, I found would be impracticable. All these towns, and many others around them, had suffered severely.

At the margin of the Lago Maorno, the barometer marked 27·10 inches, thermo. 42° Fahr. at 3 P.M., and the surface of the lake I find to be 2526 feet above the sea, at Naples. Crossing the piano, or basin of the lake, we ascend again, cross the Serras, of Cerzuto and Pizzuto, and on the ridge of the latter gain the first view of Moliterno, and dimly discern through the rainy atmosphere, Sarconi, Spinosa, Viggiano, Marsico Vetico, and even Montemurro, all towns, nearly or quite destroyed. We now commence to descend, by the side of the highest fork of the Fiume Sciavra, which falls into the Agri, along the east flank of a wild and grand ravine, with the torrent in the bottom, which, in its now swollen state, seems to be sweeping bodily before it, masses of the beds, of red and yellow clays and marls, and calcareous detritus, that to a depth of 30 to 60 feet form its boundaries, and conceal the formations beneath the sloping plain, the Piano of St. Martine.

At the highest point I passed upon the Serra di Cerzuto, the barometer stood at 26·65 inches, thermo. 45° at 5^h. 5^m. P.M. Naples time. The height above the sea was 2994·4 feet.

CHAPTER XVIII.

MOLITERNO.

A RAPID descent by a pretty good track, brought us to Moliterno, about an hour after dark, wearied and wet, after more than fifteen hours' walking and riding, twelve of which had been under heavy rain and wind. The Locanda here, though much shattered and in parts unsafe, was still tenantable, and I deemed myself happy in finding shelter and fire, for myself and my party.

Moliterno stands upon a low hill of hard and dense limestone, but generally without distinct evidence of bedding; in some spots to the east of the town there are indications of beds dipping to the west, with a moderate slope. The hill slopes rapidly towards the Sciavra, upon the south of the town, where are the ancient mills from which it derives its name. Monte Spagnoletto stands due north, and a little to the N.W. of it, distant about $2\frac{1}{2}$ Italian miles, while round thence to due west are the ridges of the lower Serras that shut out the high table land and basin of Lago Maoruo. The town, situated in the midst of rich valleys, seems a thriving place. Several large modern buildings have suffered but little, and there are some in progress of erection. The people *here* show no lack of energy in clearing away the effects of the earthquake, which, however, has dealt very mercifully with them, in comparison with towns not five miles away. At the Locanda, situated in the higher part of the town, the barometer at 8·30, Naples time, 17th Feb. reads 27·05 inches, thermo. 51° Fahr.: it is 2696·7 feet above the sea. The highest point of the Colline on which the Castello is situated is about 200 feet above this, or nearly 2900 feet the summit.

In the new buildings in progress here, the causes of such facile destruction by shock as I have remarked, are patent; no thorough bond, thick and heavy walls of ill-constructed rubble, floor beams a yard apart, inserted only 9 inches into the walls, without "tossals" or wall-plates, want of all mutual connection between the walls, floors, and roof, and both the latter of prodigious weight.

The Chiesa Madre has its axial line nearly north and south, built of brick, with limestone pilasters, &c., and brick vaulted roof, about 160 feet by 45 feet wide and 60 feet high to the springing of the vault. The transepts are vaulted also. The arch is semicircular, and is fissured widely, down to the springing, and open an inch in places. The fissures in the walls, are fine and narrow, but give good indications; general direction of wave-path, from 140° to 145° E. of north, and the slope with vertical 10° to 12°, giving $e = 11°$ from the N.W.

Nothing to indicate velocity. All around the church, are many ordinal buildings, fissured and partially thrown, and one large cardinal one, all of which indicated more or less distinctly, a wave-path from some point to the west of north to south; there were evidences, but obscure, of a minor and nearly orthogonal wave.

The Castello, consisting of heavy old massive masonry, much indurated, is severely fissured. General direction, 154° or 153° 30′ E. of north, but the extreme limits are very wide here, viz., from 105° to 165° E. of north. I have not taken a mean, but a *choice*, of those fissures whose direction appeared least likely to have been influenced by curvature (as in the great round towers) or other disturbing conditions. The fissures in the great towers of 80 feet in height or thereabouts, were open 2½ inches at the top. The walls were very thick, and such that this width, is a rude indication of the amplitude (horizontal) of the wave here.

The Chiesa, della Santa Dominica della Rosario, has its axial line 120° E. of north, or not far from cardinal: it is about 160 feet long and 50 feet wide, with a brick semi-cylindrical vaulted roof, 40 feet to the soffit, which, with the walls, are heavily fissured: the mean of seven pairs of these, gives a general wave-path of 155° E. of north, and an emergence from the north of 13° 30', the extreme limits being 10° and 17°. In this building there is evidence also in the vault fissures, of a subordinate shock, nearly at right angles to the main one.

On altars, both at the north and south sides of the church, I found wood gilded candlesticks, that had been thrown out of plumb. Those on the north side had been thrown towards the N.W. at various angles, and still leaned against the back wall or shelf of the altar. They were high up and out of reach. Those on the south of the church, I found now in their usual places, but the Sacristan informed me, that all the wood candlesticks at that side, had been thrown quite off the altar, and were found scattered about the floor, and had been since replaced.

These are decisive as to *direction* of the wave here, viz., from a point W. of north towards the south. The candlesticks at the north side, thrown by the first semiphase of the wave, were limited in motion, by the wall against which they fell and leaned, having gone over too far to recover their position at the return stroke. Those at the opposite, or south side, were at the same moment thrown out of plumb forwards, or towards the front of the altar, and finding no wall to support them, fell altogether.

In the Caffè of Gaetano Mallione, a number of bottles of the form of foreign wine bottles, full to the corks of Rosolio, stood upon a shelf, at 8 feet high from the floor, running along the west side of a wall, whose length was N. 155° east. The owner, an intelligent fellow, replaced the bottles for me in the position in which he stated he had found them in the morning after the shock. (Sketch No. 243, Coll. Roy. Soc.) Those towards the back of the shelf, (which was about 15 inches wide), leaned back against the wall, and against each other, sloping towards the east and south. Several that had stood upright close to the edge of the shelf, (which had a little ledge or curb of about three-quarters of an inch high, rising at its front edge), had been pitched over and thrown upon the floor and broken. The spots upon which these had landed, I found were on the average, three feet horizontally from where the bottles had stood on the shelf. The direction of throw was about 125° E. of north towards the N.W.; omitting one, thrown in the apparent direction x to y.

If we assume the emergence to have been 15° here, we have
$$e = 15°. \quad a = 3. \quad b = 8.$$
and the normal velocity of the wave here given by the equation
$$V = \sec e \sqrt{\frac{a^2}{b - a \tan e} \times \frac{g}{2}}$$

is 10·8 feet per second, neglecting the velocity necessary to *upset* the bottle from its base, which should be added, and would increase the velocity per second a foot or two. We cannot rely upon this for more, than a general indication, that the velocity here was not materially different from further west and north, and I could get no better data at this town.

I could obtain but very confused accounts here, of the second shock, (that of an hour or so after the great one,) many persons saying they did not feel any such, nor could I obtain any good account of the sounds; all agreed that they heard the "rombo." I was unfortunate, however, in not being able to find the Syndic or Judice, both of whom were absent.

CHAPTER XIX.

SARCONI.

I LEFT Moliterno at noon for Sarconi, still in heavy rain, with a cold N.E. wind. This town, a place of extreme antiquity and probably of Greek origin, is not above two Italian miles, in a right line nearly, east of the former, crossing the valley of the Sciavra. It stands on the lowest level of the piano, probably 700 feet, if not more, below Moliterno, upon the very edge of the steep and lofty bank of about 100 feet in depth, of deep alluvium and clays, overhanging the north bank of the Moglia, here in winter a large river, whose stony bed is at least 700 feet wide. It is spread out at the town upon horizontal beds of green and grey thin-bedded marls, with calcareous breccias and deep clays above. (See Geolog. section, Diagram No. 242, and Sketch, section, No. 244.)

Directly to the N.E. side of the town, a low colline of limestone rock rises to the height of about 350 feet above it, the hard and nearly horizontal beds of which, pierce up steep and abruptly through the clays, &c. This colline bears round the town a good way, east and north. (See section, Sketch No. 245.) The opposite, or right bank of the Moglia shows limestone in highly inclined beds in some places, and covered with deep alluvium.

Photog. No. 246 (Coll. Roy. Soc.) gives a good notion of the position of the town, and that No. 247, (Coll. Roy.

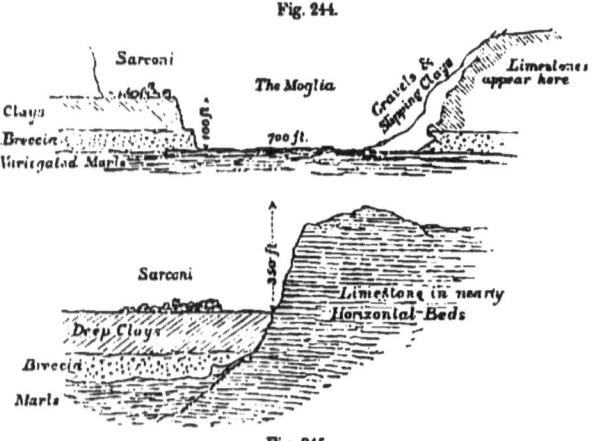

Fig. 244.

Fig. 245.

Soc.) of the character of the horizontally bedded limestone found hereabouts, as well as that of the right bank of the Moglia. The town is situated on the tongue, between the Sciavra and the Moglia, and from the general direction of wave-path hereabouts, the blow must have been delivered to it, diagonally transverse to the tongue of land, and from the limestone masses to the east and north, and was necessarily severe. From this, from the antiquity of very many of the buildings, and their ill construction, their ruin is great. The fallen houses and hills of debris are perfectly unintelligible *en masse*, but two or three isolated buildings only, present fractures or fissures. These gave an average of direction, 175° E. of north, but very unsatisfactory, both from their character, and from the high wind and driving preventing good measurement.

SAPONARA
Exterior of the Church.

SAPONARA
Remains of the Church.

The old church, however, gave better results. It was a building of great antiquity; the axial line was 55° W. of north, and, on entering the western door, I observed that two ancient and probably Roman altars, with nearly effaced inscriptions, had been built into the walls, and formed the pedestals for the stone jambs of the doorway. The building has wholly fallen in, and in great part the walls are down. The belfry tower stood at the north quoin (Photog. No. 248), a tower, of about 20 feet square, of which about 47 feet in height remain standing. The whole of the upper part has been thrown down, and the mass has fallen, partly within and partly without the church, and some on the highest remaining floor of the tower itself, all falling towards the S.E. There was a single bell in the tower, of about six cwt., which hung, as the "Parrochiano" (an aged priest, who politely came out in the rain to give me information, and whose name, I regret to find, I omitted to note) informed me, to a beam in the centre of the tower, and at

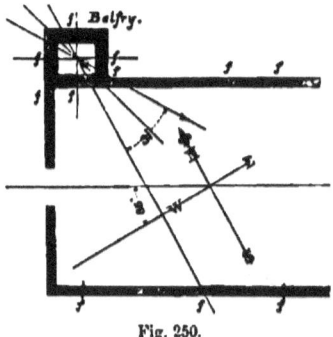

Fig. 250.

a height of 30 palmi above where I found it; namely, lying upon the top remaining floor, amidst rubbish and fallen timbers, with its mouth facing S.E. The centre of gravity of the bell, was 7 feet horizontally, distant from the centre of the tower, and the direction of its throw from the centre of the tower was 149° E. of north (Fig. No. 250).

This direction, I found coincided closely with that given by fissures, Photog. No. 249 (Coll. Roy. Soc.), two pairs of which gave a wave-path of 145° E. of north, and an angle of emergence from the N.W. = 16° 25′. This angle, as given by the isolated buildings before adverted to, appeared to be but 11° to 11° 30′; I rely, however, upon those of the church.

The bell had fallen vertically 30 palmi $= 30 \times 0\cdot 862$ ft. = nearly 26 feet English. We have, therefore,

$$e = 16°\cdot 25' \qquad a = 7 \qquad b = 26,$$

and the value of

$$V = 9\cdot 78 \text{ feet per second};$$

a result necessarily below the truth, as the effort of the wave, was communicated to the bell, not through the centre of gravity, but by the pintles; so that the first effect, was merely to make it *swing*, contrary to the direction of throw. We may estimate the velocity as at least 2 feet per second more, or about 12 feet per second for the wave here, which, although not exact, sufficiently corroborates previous determinations.

Several very thick and very ancient walls here, built of small bad rubble, and showing themselves to have been very *dead* and inelastic, presented quoin fissures, open $4\frac{1}{2}$, 5, and even $6\frac{1}{2}$ inches, at 20 to 30 feet from base; the average might be taken at $4\frac{3}{4}$ to $5\frac{3}{4}$ inches, and seem to indicate a wave of long amplitude here.

CHAPTER XX.

SARCONI TO SAPONARA.

I LEFT Sarconi, where little could be learned, and pushed on by the valley of the Moglia, through about two miles of fine oak forest for Saponara.

Having recrossed the Sciavra, I paused upon the S. E. slope of a low hill, south of Saponara, and within half a mile of what was a few weeks since, a town of six thousand inhabitants, now a shapeless heap of ruin.

To the west, on the recent ruins of a Capuchin convent, under Il Monte, and to the east, on the opposite bank of the Sciavra, on the piano, in the fork between it and the Agri, are the ruins of the ancient Roman city of Grumentum.

The low hill, on which I am, like all those hereabouts, is of ill-compacted limestone, in beds nearly horizontal, dipping here 10° W., and much covered with deep alluvium. These must be reposing perfectly unconformably, against the limestone beds, of which the lofty and steep conoidal hill upon which Saponara stood, is composed.

These run through the mass of this hill, tilted to within 15° of vertical, and in a direction, generally, of a few degrees to the west of north and south.

The hill itself is nearly conoidal, rather narrower in its N. E. and S. W. diameter, than at right angles thereto: it is extremely steep, and rises from 900 to 1000 feet (by the eye) above the river Agri.

The town covered the summit, and spread some way down the flanks all round, descending most, upon the east side. The ancient Norman Castello Cilliberti, crowned the crest, which was bare limestone rock, and equally bare for 200 or 300 feet downwards. Beds of increasing thickness of alluvium then begin to cover it, and as it slopes down rapidly at an angle of nearly 45° to the Agri, these become 70 to 100 feet thick.

The opposite bank of the river, running N. W. to S. E., which I can see about 400 feet below me, consists of still deeper alluvium, forming part of the Piano Spineto, &c., resting upon horizontal and exposed beds of clays and schaly argillaceous rocks.

To the north and N. W., beyond the hill of Saponara, the Piano of Mattine delle Rose, extends for some five or six miles, all of such formations, and beyond that the mountains rise to Marsico Vetico, and beyond to the crest of Monte Voltorino. It was from this direction that the blow reached Saponara—delivered end on through its vertical beds, from the vast mass of loose material of the piano. Insulated to a depth of 1000 feet, the Agri and Sciavra running round its base from N. W. to south, peaked, narrow, and abrupt, and surrounded by horizontally abutting, dense and inelastic formations, the summit of the hill and the unhappy town upon it, must have shook and swayed like a mast, after the shock of the 16th of December.

The appearance of the ruin of Saponara was appalling.

SARAGOSSA
WITH THE REMAINS OF THE CASTILLO ALJAFERÍA

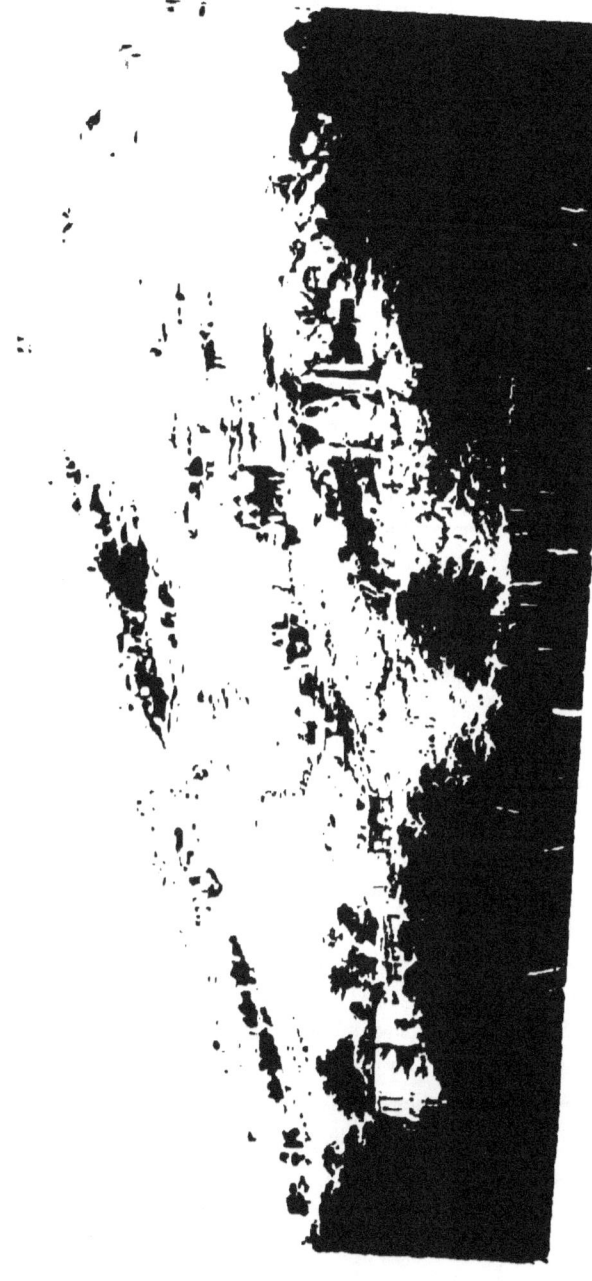

THE MOUND OF KUHOSH WHERE SAMSNARA HAD BEEN.
Looking Southward

As I advanced, and gradually ascended close to it upon the S. E., and looking north, literally nothing remained standing, upon the crest, but the Castello Cilliberti (Photog. No. 251). Its roofs and floors were fallen in, its towers split or fallen, and many of its massive, indurated, ancient, and buttressed walls prostrated. All the upper part of the town, that which had been the most thickly built upon and densely inhabited, was now strewed around in featureless ruin; acres of ground were covered and heaped, with rounded mounds and sloping avalanches of white and dusty debris, but not a wall, or even the base of one, standing or visible.

Upon the very summit (to the right of the Castello Cilliberti, Photog. No. 251) may be seen projecting above the mass of rubbish the solitary fragment remaining erect of the ancient church,—a building of enormous solidity, massive, and ancient—and whose masonry, indurated by time, attested, by the enormous blocks in which some of its quoins were dislodged, its resistance to the terrible violence of the shock here. These are seen from a closer point of view in Photog. No. 252, looking nearly northward.

It was not, however, until having passed beneath the town, to the north of it, and looked back upon it, that the awful character of its desolation became fully developed.

Seen from this side (Photog. No. 253) the summit and far down the slope all round, presented nothing but a rounded knoll—shadowless and pale—of chalky stone and rubbish, without line or trace of street or house remaining; it might have seemed an abandoned stone quarry, or the rubbish of a chalk pit, save that its rounded and monotonous outline, was broken here and there by beams and blackened timbers that rooted in the rubbish stood thrown up in wild confusion against the sky like the gaunt arms of despair.* Such an horizon high above me, and close around, on all sides the cold and dripping huts, thrown together, of whatever ruin 'had first presented to the hand, and filled with wounded and perishing beings, depressed the spirits of even the passing observer, to an extent that made me readily comprehend, the dull and stupefied patience, with which the survivors still cowered round the ruin of their homes.

There was less of the town upon the north, than upon the south slope of the hill; and as the blow came from the N. W., the "free lying stratum" was at the south and S. E. side, so that the condition for destruction, and the extent of food for it, were both here the greatest possible.

The only portion in which walls were standing at all, and still showing something of the features of a town, though shattered and in great part roofless, was at the east side, a long way down from the summit, as seen in part in Photog. No. 251, and in Photog. No. 254 (Col. Roy. Soc.), from the east side, looking back westward and northward.

No more complete proof could have been afforded of the fact, that the utter destruction of the town was due to the swaying of the hill itself upon which it stood, and not to some great increase at this region, of the dimensions or velocity of the earth-wave itself, than the finding several buildings around the *base* of the hill, and upon the deep alluvium close to its junction with the limestone of which it is composed, *comparatively* safe—all, however, severely shattered. Of these an example occurs in a house of two stories, seen to the left and low down in Photog No. 253.

---

* These timbers had been removed for huts and fuel between the time of my visit and that of taking the photographs, February to May, 1857.

The Castello, and the portion of the town below it to the east, gave abundant measures, of direction and emergence. The wave-path was sub-abnormal to almost every building in the place, and vast wedge-shaped masses were thrown out everywhere, some of which may be seen in the Photographs.

The examination of the buildings of the Castello resulted in a wave-path 150° E. of north to south, and an emergence of 14° to 16° from the north.

The buildings on the east slope gave a rather different wave-path; the average of a great many giving 120° 30' E. of north, and a less emergence 12° to 15° from the north.

I take the former, however, as the true wave-path here, and deem the difference lower down to arise from the gyratory oscillation of the hill itself. A further proof of this was, that on traversing round its base, I found that while all the buildings there situated, gave a *prevalent* direction of wave-path about the same as above, they also showed complicated secondary fissures in directions that indicated their production by oscillations, emanating everywhere radially from the centre of the hill. I conclude, therefore, that the first great blow which prostrated in a moment the town, came as above. The hill itself, set to oscillate in the same plane as that of the path of the wave, rapidly began to oscillate in other planes, in fact, became a conical pendulum, and hence whatever buildings remained standing but fissured, by the primary blow, were again fissured in new directions or prostrated by this proper motion of the mass of the hill.

There were even evidences in the Castello at top (where these secondary motions were, of course, a maximum) of portions that had been broken out and disjointed, in the direction of the original wave, and by it, having been thrown afterwards, in directions nearly orthogonal to it.

It was difficult to get any indication, from which to calculate velocity here, or any approximation to the amplitude of the wave, nearly everything being totally prostrate and the history of its fall, a blank; the bells, for example, of the church, were many feet deep under stone and rubbish.

The direction of the wave being generally sub-abnormal, rendered it difficult to find any walls at the Castello, directly across the plane of which, the force had passed, and so circumstanced, that, being found still standing, they would afford to calculation, a limit beyond which, the *total* velocity of the summit of the hill could not have reached. In fact, had the wave and first oscillation of the hill, together been normal or subnormal to the walls of the Castello, not one stone of it would have been left standing; its remnant owed its remaining erect, mainly to the diagonal direction of its walls in reference to the wave, as may be readily seen, in the case of the buttressed curtain wall to the left in Photog. No. 251.

I was enabled, however, to find one massive piece of wall at the north side, whose condition admitted of its being used as a tolerable measure of this limiting velocity. Its lengthway was 72° E. of north; it was therefore within 12° of being transverse to the path of the primary wave.

It was a curtain wall of old rubble masonry, connected with buildings or towers at both ends about 40 feet apart, but in advance of one, so that it had very slight bond with that tower; while at the other an old settlement, or crack, broke a good deal of the connection. This wall was cracked nearly horizontally, about a foot above its base, and leaned over to the north about 1½ inch. It had no batter: its thickness was 2·75 feet, and its height 20 feet. We can calculate the horizontal velocity necessary to have fractured its base, without overthrowing it, from the equation—

$$V = \tfrac{2}{3} g \times \frac{b \, L}{a^2}.$$

here $b = 2\cdot75$ feet; and to allow for the wall being 12° out of square to the wave-path, we shall take it $= 2\cdot8$ feet; $a = 20$ feet, $L = 2 \times 52$ feet. The coefficient for rubble limestone masonry of this highly indurated mortar being fully double that of ordinary work. Then

$$V = 21\cdot46 \times \frac{104 \times 2\cdot8}{400} = 15\cdot623 \text{ feet per second.}$$

I have no measure of the value of e, for Saponara, except those of the Castello, the maximum of which was 16°. We may assume it about the same as at Sarconi, $= 16° \, 25'$; and taking the velocity of the wave itself (in its normal direction) to be the same as at Polla $= 12\cdot97$ feet per second, we have its *horizontal* velocity $= \dfrac{12\cdot97}{\sec e} = 12\cdot45$ ft. per second; deducting which from the preceding, we obtain 3·173 feet per second as the velocity of *oscillation* of the hill of Saponara itself, which may be considered as having been horizontal in direction.

The extreme limit of total velocity then, at Saponara, cannot have exceeded about sixteen feet per second, only about three feet per second greater than at Polla, Padula, &c.—a striking proof how small an increment of velocity, is sufficient to sweep all before it.

CHAPTER XXI.

SAPONARA TO SPINOSA, AND THE ENTRANCE OF THE VALLEY OF THE LADERANA.

Leaving Saponara, where there were no authorities but a few gendarmes to be found, one of whom I took on with me to Montemurro, which I proposed to reach late, (as a guide rather than a protection against marauding on the part of the starving people hereabouts,) I rode along the right bank of the Agri eastward, to near the junction with the Moglia, and then forded the former with some difficulty, owing to its swollen state; and after half a mile again forded the Aqua Fredda, or Frigida, which, coming down by the little vallone of the same name, from the heights of Monte dell' Agresto, and the Santo Spirito, falls into the Agri on its left bank. Looking back at Saponara, from 2 miles distant, to the east, and after crossing these rivers again at about 4 miles, I made the sketches (Nos. 255 and 256) of its appearance and relation to the hills around, and to the piano, &c. The afforested low hill in front, is the Bosco di Guardia Maura, and part of the Bosco dell' Aspro. The Agri, with deep and precipitous clay banks, flows between Saponara and the observer.

It is scarcely conceivable that Saponara will ever be rebuilt, the destruction is too absolute, to leave sufficient

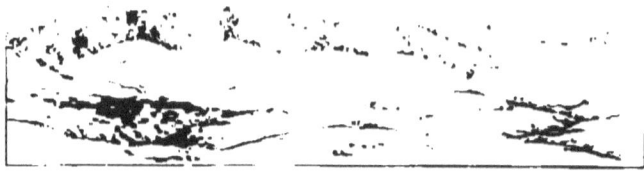

SAPONARA.

Sketch Pl. 256

SAPONARA.

The Beds of the AGRI & MOGLIA.
Denudation of the Strata.

inducement to remove the mountainous masses of rubble and rubbish, that must form the necessary preliminary. Those associated with the place will find another site, and rekindle their hearths on strange ground, from which their surviving successors, will within another century most probably be driven forth, by a future great earthquake, from houses as unskilfully constructed, as those their sires perished beneath.

. As I looked back once more upon the place, I came to understand that thus it has been, that we find in Southern Italy, such numbers of old and new towns of the same name, situated not far apart—such as Corneto, Vecchia e Nova, Tito, Vecchio e Nova, Capaccio, Marsico, and numbers of others—and it recurred to memory that after the great shock of 1783, several of the Calabrian towns, were then rebuilt on new sites—as St. Agatha, Vecchia e Nova, Bianco, Vecchia e Nova, &c.

Sir Charles Lyell's vivid sketch, of the probability of future ages, finding in the formations of to-day, human remains and objects, and records of human art, ('Elements of Geology,' chap. xlvi.), embedded and preserved, too, suggested thoughts of the future state and after history, of the mounds that are alone left where Saponara was.

Beneath these masses lie, some few mangled human remains, that will never probably be found for sepulture—those of domestic animals, and of the mammalian and other vermin that follow man—fragments of every household utensil, personal and domestic ornaments, weapons, tools, and instruments, carved and wrought stone, ivory, hard wood, grain, fruits, food of many sorts, books, records, crumpled pictures, glass and pottery, wrought timber in the splintered joinery, as well as the metallic parts of houses, &c.; under what changed conditions as to geological position may these, if ever, see the light again!

In this vitalizing climate, but a few years will elapse, before the frosts of winter and its torrential rains, will have pulverized and reduced to interstitial soil, much of the dry rubbish: plants and seedlings will root in it; forest trees will send their searching roots downwards through it, and beneath their fostering shade, a jungle of shrubs, weeds, mosses, and coarse grass, will spring up and conceal with verdure the harsh colouring and form, of the heaps that once were a city; and after but a few generations the fearful fate of its two thousand overwhelmed inhabitants—nay, its very site, will have become a tradition as dim, as that of the neighbouring Grumentum.

About half a mile S.E. of Saponara, upon the level clays of the piano, I passed two square gate-piers of rubble ashlar masonry, leading to an orchard, 3 feet square, and 7 feet in height, both prostrated, and in directions accurately parallel, 140° 30′ E. of north, fractured at the ground level from their foundations. The mortar was bad, and by examination with the hand I judged had not an adhesion

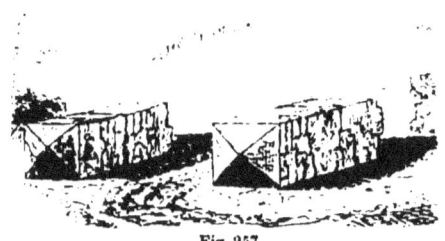

Fig. 257.

of more than about 2 lbs. per square inch. The wave-path was exactly subnormal to the piers (Fig. 257).

The horizontal velocity for fracture from the equation

$$V = \tfrac{2}{3} g \times \frac{b\,L}{a^2} \text{ is therefore}$$

$$V = 5\cdot 48 \text{ feet per second.}$$

The horizontal velocity for overthrow after fracture from the equation

$$V^2 = \tfrac{2}{3} g \times \sqrt{a^2 + b^2} \times \left(\frac{1 - \cos\theta}{\cos^2\theta}\right)$$

θ being $22°\ 40'$, $a = 7$ feet, $b = 3$ feet, is

$$V = 5\cdot 14 \text{ feet per second:}$$

The total horizontal velocity for fracture and overthrow is therefore

$$V = 5\cdot 48 + 5\cdot 14 = 10\cdot 62 \text{ feet per second:}$$

but $e = 16°\ 25'$ assumed the same as at Sarconi; therefore

$$V = 10\cdot 62 \times \sec e = 11\cdot 04 \text{ feet per second,}$$

the actual velocity of the wave in its direct path. This result—deduced from *two* similar blocks of masonry, of the simplest and best form, natural seismometers in fact, and coinciding so closely with previous and distant determinations—affords a strong confirmation of the correctness of the explanation given, of the nature of the higher velocity that overthrew Saponara, close to which, but down on the level of the plain, we see thus the wave has its ordinary velocity.

I pursued my way towards Montemurro, where I hoped to rest.

At 4 miles S. E. of Saponara, under Spinosa, on the level of the bed of the Agri, at $3^h\ 15^m$ Nap. time, the barometer read $29\cdot 11$ inches, thermo. $42°$, and gives the height above the sea $= 649\cdot 1$ feet.

This may be viewed as about the lowest level of the Piano Mattine, so that its mean level is nearly the same, as that of the great Piano of Diano.

For some miles about the junction of the Moglia and Agri, evidences of prodigious river erosion exist. In many places in the main river-bed—here from 500 to 700 feet, or upwards, in width, though rarely covered wholly with water—great insular masses challenge astonishment, by the rate at which they are being carried bodily off in winter.

One of the more remarkable of these is seen in Sketch No. 258: they all consist of great beds of calcareous breccia, resting conformably upon perfectly level deep beds of extreme thinness of parallel lamination of green, grey, and purple clays, or marls, hard, dense, and unctuous, but rapidly softened and dissolved when wetted. Above the breccia, lies an immense thickness of dense red brown, clay and loam; the laminated marl beds exist, just at the level of the watercourse of the two rivers, and as these get rapidly sapped and cut away, huge masses of the breccia, break off and fall separate into small pieces, and with the clays shed off their summits, are swept away, leaving nothing deposited finally upon the river-beds, but the harder calcareous pebbles and boulders of the breccia, and those still harder travelled boulders, which it contained in abundance. Amongst the latter, are many of sienite and yellow granites, some of a green and white fine sandstone breccia, having lithologically a most suspicious look of indurated chalk, from a green sand formation, and many of variegated jaspers. Large blocks of the latter, banded with green and purple, are found abundantly in the clays and on the slopes to the north side of the valley, and are plainly the product of intense metamorphic action, upon the variegated marl-beds.

Passing beneath Spinosa, I scanned the town narrowly with the telescope; but although many buildings were prostrate and fractured, it did not appear to offer much, to reward the time and labour of ascent; and being only about five miles in a right line from Saponara, ascertainment of direction here was of less importance. I obtained from some houses at the base, however, a satisfactory measurement of the latter from fissures, which gave 134° 30' E. of north for the wave-path. Owing to circumstances of stone-work and apertures, &c., they could not be relied upon as to emergence, beyond proving that it was from the N.W. The owner of one of those houses informed me, that his business frequently brought him to Castel Saraceno, about ten miles to the south, and that there the shock had been felt from N. W. to S. E., or 135° E. of north.

CHAPTER XXII.

JOURNEY TO MONTEMURRO.

WE passed the Agri again, narrowly escaping the loss of one of the laden mules, owing to the large stones in the bed, the torrent of muddy water taking them above the girths; and commenced a long and toilsome ascent, along the small lateral valley of the Fiume Levada, or Laderana, crossing it several times, from the west to the east bank.

This stream is not named on the maps of Zannoni, or of Bachler d'Albe, and no two people hardly, seemed to pronounce its name quite alike. Once landed on the left (north) bank of the Agri, we reach a new set of formations. The limestone and breccia here disappear, and are succeeded by thick argillaceous beds, with thin bands of something approaching to clay iron-stone; some beds of calcareous clays, much indurated, and occasionally, heavy beds of a yellow and grey calcareous soft sandstone, all not very far from horizontal, and dipping to the north and N. E. The whole of these are overlaid, by enormous deposits, of dense tenacious clays, red, brown, and yellow, almost without a pebble. These stand, as soft shedding cliffs, above this stream—now a brawling torrent of liquid mud, which is undercutting and sweeping them away, in masses. In many places 400 feet in depth of these clays, overlay the soft rocks beneath. Near the junction of this torrent with the Agri, I had noticed many fragments, and some large lumps of lignite in its bed, and when ascended, to within about a mile of Montemurro, I was enabled to see the lignite beds *in situ* beneath the clay cliffs at the opposite (east) side of the Laderana, and nearly on the level of the water. They appeared to be from 1 to 2 feet in thickness, perfectly black, but as fuel, of very inferior quality; they are unused, and apparently unknown to the inhabitants.

A good while before reaching this elevation, Gallichio, Missanello, and other distant towns, to the east and S. E., perched amidst the lateral valley recesses of the Agri, had been visible, all showing with the telescope, evidences of devastation. St. Archangelo is the most remote, that I have had a glimpse of, distant about 20 miles to the S. E.

END OF VOL. I.

www.ingramcontent.com/pod-product-compliance
Lightning Source LLC
Chambersburg PA
CBHW021424300426
44114CB00010B/642